REPORT

Evaluating the Desirability of Navy F/A-18E/F Service Life Extension Programs (SLEPs)

Edward G. Keating, Irv Blickstein, Michael Boito,
Jess Chandler, Deborah Peetz

Prepared for the United States Navy

NATIONAL DEFENSE RESEARCH INSTITUTE

The research described in this report was prepared for the United States Navy. The research was conducted within the RAND National Defense Research Institute, a federally funded research and development center sponsored by the Office of the Secretary of Defense, the Joint Staff, the Unified Combatant Commands, the Navy, the Marine Corps, the defense agencies, and the defense Intelligence Community under Contract W74V8H-06-C-0002.

Library of Congress Cataloging-in-Publication Data is available for this publication.

ISBN: 978-0-8330-5104-2

Published 2010 by the RAND Corporation
1776 Main Street, P.O. Box 2138, Santa Monica, CA 90407-2138
1200 South Hayes Street, Arlington, VA 22202-5050
4570 Fifth Avenue, Suite 600, Pittsburgh, PA 15213-2665
RAND URL: http://www.rand.org/
To order RAND documents or to obtain additional information, contact
Distribution Services: Telephone: (310) 451-7002;
Fax: (310) 451-6915; Email: order@rand.org

Preface

In the spring of 2009, Arthur Barber, Deputy Director, Systems Analysis Division, N81B, Office of the Chief of Naval Operations, asked the RAND Corporation to undertake a study titled "Estimating Optimal Management of Aging Aircraft in the Department of Navy." The objective of this project was to assist the Navy and Marines in making optimal decisions with respect to their aging aircraft. As Department of Navy aircraft age, leaders face decisions, such as whether to modify and upgrade the aging systems or whether to replace them. The Department of the Navy wants to make such aging-aircraft management decisions in an objective and analytical manner that provides the most military efficacy to the department for a given level of spending (or costs the department and the taxpayer the least for a given level of military efficacy).

The Navy asked RAND to first investigate service life management of the F/A-18E/F fighter jet. While the F/A-18E/F is relatively new, preliminary planning has begun as to the feasibility and desirability of a service life extension program (SLEP) on the aircraft. This study complements ongoing Navy examination of E/F SLEP options by setting forth a methodology to compare the value of doing an E/F SLEP to the alternative of buying replacement Joint Strike Fighters (JSFs).

Given the many parameters that are unknown, this study does not present any final conclusions about E/F SLEP desirability. Instead, the study focuses on different methodologies. For each methodology, we present ranges of parameter values that favor versus oppose undertaking E/F SLEPs.

The Navy subsequently asked RAND to analyze C-2A issues using the same methodologies. These findings are presented in Appendix A.

This research should be of interest to Navy and other Department of Defense personnel involved with aviation acquisition and maintenance issues.

This research was sponsored by the U.S. Navy's Systems Analysis Division, N81B, and conducted within the Acquisition and Technology Policy Center of the RAND National Defense Research Institute, a federally funded research and development center sponsored by the Office of the Secretary of Defense, the Joint Staff, the Unified Combatant Commands, the Navy, the Marine Corps, the defense agencies, and the defense Intelligence Community.

For more information on the RAND Acquisition and Technology Policy Center, see http://www.rand.org/nsrd/about/atp.html or contact the director (contact information is provided on the web page).

Contents

Figures

Tables

Summary

This report resulted from a U.S. Navy request to the RAND Corporation to assess the cost-effectiveness of prospective service life extension programs (SLEPs) on F/A-18E/F fighter jets and C-2A transport aircraft. The Navy's request and the methodology we present in this report assume that a SLEPed aircraft and the potential replacement aircraft have comparable capabilities.

SLEPs are complex depot-level overhauls in which an aircraft is extensively inspected, cracked or worn materials are repaired or replaced, computer systems are upgraded, and mechanical function is ensured. When a SLEP is undertaken, it is assumed that a number of extra years of operation are added to an aircraft's life. Likewise, one might hope that a post-SLEP aircraft performs better, e.g., has greater availability levels or lower maintenance costs.

However, many key parameters are uncertain, e.g., the cost of the SLEP, the additional years of service provided by the SLEP, the cost of the alternative new aircraft. Given this uncertainty, we assess which parameters are most crucial to decisionmaking and present ranges of their values that favor versus oppose undertaking SLEPs.

Making the Cost-Minimizing Choice Between a Service Life Extension Program and a New Aircraft

One can estimate the discounted sum of total costs associated with a SLEP versus a new aircraft. However, these summations are not directly comparable, since a new aircraft will presumably last longer than the number of extra years of operation provided by a SLEP.

One way to correct for this problem would be to annuitize each approach's cash flows and to choose the approach (SLEP versus new aircraft) with the lower annuitized cost. Annuitization translates uneven cash flows (e.g., an up-front procurement or SLEP payment followed by lower year-to-year maintenance costs) into a single, equal-sized annual payment. Annuitization is appropriate if technology is static, i.e., the Navy's options upon the expiration of the once-SLEPed aircraft are neither better nor worse than they are today.

Unfortunately, choosing the approach with the lower annuitized cost does not consider aircraft availability issues.

Consideration of Aircraft Availability Trends

As aircraft age, we expect their annual maintenance costs to increase and their annual availability levels (the fraction of the fleet that is operator possessed and mission capable) to decrease. Selection of the cost-minimizing option between SLEP and aircraft replacement ignores prospective aircraft availability declines.

Keating and Dixon (2003) suggested use of average cost per available year to account for age-driven availability declines. While the average-cost-per-available-year metric is tractable and intuitive in how it adjusts for changing availability, it is not a standard, agreed-upon metric. One must accept a priori that average cost per available year is the right metric to minimize.

A different methodology would be to assume that Navy valuation of an aircraft increases linearly in its availability level, i.e., net benefit equals $k \times a(t) - c(t)$, where $a(t)$ denotes availability in year t and $c(t)$ denotes cost in year t. One can then annuitize net benefits of both new aircraft and a SLEP and choose the option with the greater annuitized net benefit.

Introduction of an availability valuation parameter k has its own challenges, however. We can only bound, not specifically identify, k.

If we assume that the new aircraft has net benefit, the new aircraft's k must be greater than the new aircraft's average cost per available year. On the other hand, the new aircraft's k must be small enough that it is optimal to operate the new aircraft for its assumed lifetime (and not retire it sooner). The result of these two constraints is a range of possible k values. How should one interpret, for instance, a SLEP being worthwhile if k is at its minimum value but not if k is at its maximum value? There is no clear way to assess the Navy's value of k.

F/A-18E/F Context

The Navy asked RAND to evaluate the desirability of a prospective SLEP on F/A-18E/F fighter aircraft. F/A-18E/Fs will be due for a SLEP or for retirement sometime late in the 2010 decade or early in the 2020 decade.

Our models need an estimate of the cost of the SLEP, as well as the additional years of operation it allows. One also wants to know the life-cycle costs associated with the replacement aircraft alternative to doing an E/F SLEP. If aircraft availability trends are to be considered, one would also need to estimate the levels of aircraft availability associated with both a post-SLEP F/A-18E/F and its prospective replacement.

Unfortunately, an F/A-18E/F service life assessment program only recently commenced. So, many of the parameters we need are not known and can only be estimated with a high degree of imprecision.

Based on insights from Navy experts, analysis of Navy data, and literature review, we make the following assumptions:

- The alternative to undertaking E/F SLEPs is to acquire the F-35C carrier variant of the Joint Strike Fighter (JSF).
- An E/F SLEP would add ten years to the aircraft's life. (We refer to a SLEP that adds ten years to an existing aircraft's life as a *ten-year SLEP.*)
- This SLEP would cost \$26 million (fiscal year [FY] 2009 dollars) per aircraft.

- F/A-18E/F maintenance and modification expenditures per aircraft will grow at 3 percent per annum in real terms.
- A ten-year SLEP would "pull back" annual expenditures per aircraft to the level seen when the aircraft was 11 years younger (ten years of pullback plus the year spent in the SLEP).
- F/A-18E/F aircraft availability will follow the downward trend observed in the F/A-18C/D fleet and a ten-year SLEP will return the aircraft's availability to its level seen 11 years prior.
- A JSF will have a 30-year lifetime and a procurement unit cost of $80 million.
- JSF maintenance and modification costs will be the same as an E/F's at the same age, but JSFs will not require a SLEP after 20 years of operation.
- JSF availability will equal assumed E/F availability through age 20, then stabilize.

F/A-18E/F Service Life Extension Program Desirability Analysis

Minimizing either the cost metric or the average-cost-per-available-year metric, we find that a $26 million SLEP that adds ten years to the life span of an F/A-18E/F is desirable relative to buying a new JSF.

Use of the average-cost-per-available-year metric consistently finds results more favorable to a new aircraft than use of the cost metric. With our baseline parameters, the differences are moot. Both methodologies favor the SLEP. But the average-cost-per-available-year metric, because it considers aircraft availability, has a lower SLEP cost cutoff ($30.9 million versus $33.6 million), a greater SLEP year cutoff (nine extra years versus eight), and a greater JSF procurement unit cost cutoff ($67.9 million versus $62.1 million).

Net benefit maximization also favors ten-year $26 million E/F SLEPs with our baseline parameters. But now one attains a range of SLEP cost cutoffs (between $28.2 million and $30.9 million), JSF procurement unit cost cutoffs (between $68.4 million and $74.6 million), and SLEP year cutoffs (nine or ten). These cutoff ranges emanate from uncertainty in the valuation parameter k, a parameter we can bound but not identify.

If the Navy attributes even moderately greater military value to a JSF than to a SLEPed E/F, our generally pro-SLEP findings can reverse. The Navy must therefore examine the extent to which it might put greater military value on JSFs. Such an inquiry is beyond the scope of this analysis.

Conclusions

This report applies three different methodologies to assessment of the desirability of an F/A-18E/F SLEP. With our baseline parameters, all three methodologies favor undertaking the SLEP.

Minimization of average cost per available year and net benefit maximization consider aircraft availability levels. Cost minimization does not.

While the average-cost-per-available-year metric is tractable and intuitive in how it adjusts for changing availability, it is not a standard, agreed-upon metric. One must accept a priori that average cost per available year is the right metric to minimize. The net benefit maximization approach assumes that Navy valuation of an aircraft increases linearly in $a(t)$.

We conclude that an analysis of repair-replace decisionmaking should start with cost-minimization calculations. Most notably, if a new aircraft is found to be less costly, there is little need for additional calculations assuming the typical result of newer aircraft having greater availability and capability levels. If, on the other hand, the repair approach is found to be less costly, we then recommend explorations of both net benefit maximization and average-cost-per-available-year minimization to see how much additional availability or capability and the value placed on them it would take to offset the higher cost in decisionmaking.

Acknowledgments

We appreciate the research sponsorship of Arthur Barber, Deputy Director, Systems Analysis Division, N81B, Office of the Chief of Naval Operations.

LCDR Scott Russell was a very involved and enthusiastic point of contact. We are most appreciative of his extra efforts on our behalf. We also appreciate the assistance of CAPT Eric Kaniut, CDR Michael Donnelly, CDR Fernando Maldonado, CDR Kevin Sandlin, Kevin McCarthy, and Todd Standard.

This research was briefed to RADM Brian Prindle on March 17, 2010.

RAND benefited from an informative visit to the Patuxent River Naval Air Station and the F/A-18–related insights and assistance there of Patricia Behel, Tim Conley, R. L. Gage, Richard Ryan, and Laurence Stoll.

On separate visits to Patuxent River, RAND received C-2A insights from LCDR Harry Jaeger, Dennis Heeren, Joanne Heilmeier, Grace Henderson, Eric Johnsen, Angie Knappenberger, Randy Lefler, Carolyn Ruthenberg, and Jason Thomas.

Brent Boning of the Center for Naval Analyses shared a number of valuable insights with the RAND research team.

We are appreciative of the insights we derived from the Aircraft Inventory and Readiness Reporting System (AIRRS) and Decision Knowledge Programming for Logistics Analysis and Technical Evaluation (DECKPLATE), two data systems provided by the Naval Air Systems Command (NAVAIR). The NAVAIR personnel who run the Naval Aviation Logistics Data Analysis systems, such as AIRRS and DECKPLATE, have a positive customer service attitude that impressed us considerably.

We received constructive reviews of earlier versions of this report from Michael Alles of Rutgers University and Frank Camm and Yool Kim of RAND. We thank Camm for suggesting Appendix B.

RAND program director Philip Antón provided helpful comments on earlier versions of this report. Eric Peltz also provided comments on this report. Mark Arena and Vera Juhasz assisted with the document review process. Lisa Bernard edited this report. Rick Bancroft, Le Roy O. Gates, and Ravi Rajan provided valuable computer assistance. Benson Wong helped prepare the document.

Of course, research findings, interpretations, and remaining errors are solely the authors' responsibility.

Abbreviations

AIRRS	Aircraft Inventory and Readiness Reporting System
COD	carrier onboard delivery
DECKPLATE	Decision Knowledge Programming for Logistics Analysis and Technical Evaluation
DoD	U.S. Department of Defense
EIS	equipment in service
FY	fiscal year
GDP	gross domestic product
JSF	Joint Strike Fighter
MC	mission capability
NAVAIR	Naval Air Systems Command
O&S	operating and support
OMB	Office of Management and Budget
SLEP	service life extension program
VAMOSC	Visibility and Management of Operating and Support Cost

CHAPTER ONE

Introduction

This report resulted from a U.S. Navy request to the RAND Corporation to assess the cost-effectiveness of prospective service life extension programs (SLEPs) on F/A-18E/F fighter jets and C-2A transport aircraft. The Navy's request and the methodology we present in this report assume that a SLEPed aircraft and the potential replacement aircraft have comparable capabilities.

SLEPs are complex depot-level overhauls in which an aircraft is extensively inspected, cracked or worn materials are repaired or replaced, computer systems are upgraded, and mechanical function is ensured. Aircraft undergoing SLEPs are expected to be out of service for a number of months. SLEPs are worked through a fleet of aircraft over time, so the process for an entire fleet would likely extend over a number of years. SLEPs can be expensive, but so too is replacement of a fleet of aircraft, the alternative to which a SLEP is compared.

When a SLEP is undertaken, it is assumed that a number of extra years of operation are added to an aircraft's life. Not surprisingly, the number of additional years of operation added by a SLEP is an important parameter, with greater increases making a SLEP more desirable. Likewise, one might hope that a post-SLEP aircraft performs better, e.g., has greater availability levels or lower maintenance costs. Until a SLEP is undertaken on a number of aircraft, there will be uncertainty as to the additional years provided by a SLEP, as well as the post-SLEP availability and maintenance cost patterns. There is likewise uncertainty as to how a new aircraft will perform. Uncertainty is ubiquitous in aircraft decisionmaking.

Given this uncertainty, we assess which parameters are most crucial to decisionmaking and present ranges of their values that favor versus oppose undertaking SLEPs.

The remainder of this report is structured as follows: Chapter Two presents a model of cost minimization and shows how the annualized costs of different approaches can be compared. Chapter Three builds on Chapter Two, considering aircraft availability as well as costs. We present a simple way to consider aircraft availability as well as a more complex net benefit maximization methodology. Chapter Four presents and discusses data on a prospective SLEP on F/A-18E/Fs. Chapter Five then applies the F/A-18E/F data to our three different methodologies. Chapter Six concludes the report. The report also provides two appendixes. Appendix A uses our techniques to evaluate C-2A issues. Appendix B shows how our annuitization approach yields policy implications consistent with those of the net present value analysis prescribed for government cost-benefit analysis by the Office of Management and Budget (OMB).

CHAPTER TWO

Making the Cost-Minimizing Choice Between a Service Life Extension Program and a New Aircraft

In this report, we consider the Navy choosing to undertake a SLEP on an existing aircraft versus choosing to replace it with a new aircraft. If the SLEP is undertaken, we assume that it will provide Y_R additional years of service.[1] We assume that a new aircraft will provide Y_N years of service. One might expect, though it need not be the case, that $Y_N > Y_R$.

Let $c_N(t)$ denote the constant dollar costs of a new aircraft in year *t*. Then the new aircraft's life-cycle cost would be

$$\sum_{t=0}^{Y_N} \frac{c_N(t)}{(1+i)^t}, \tag{2.1}$$

where *i* is the long-term real interest rate. Likewise, the costs of a SLEPed aircraft would be

$$\sum_{t=0}^{Y_R} \frac{c_R(t)}{(1+i)^t}. \tag{2.2}$$

OMB's Circular A-94 (2002) instructs that

> the standard criterion for deciding whether a government program can be justified on economic principles is *net present value*—the discounted monetized value of expected net benefits (i.e., benefits minus costs). Net present value is computed by assigning monetary values to benefits and costs, discounting future benefits and costs using an appropriate discount rate, and subtracting the sum total of discounted costs from the sum total of discounted benefits. . . . All future benefits and costs, including nonmonetized benefits and costs, should be discounted. . . . Analyses that involve constant-dollar costs should use the real Treasury borrowing rate on marketable securities of comparable maturity to the period of analysis. This rate is computed using the Administration's economic assumptions for the budget, which are published in January of each year. A table of discount rates based on the expected interest rates for the first year of the budget forecast is presented in Appendix C of this Circular. (pp. 4, 8, 9)

[1] We refer to such a SLEP as a Y_R*-year SLEP.*

We therefore believe that i should be set equal to OMB's prescribed long-term rate, 2.7 percent in 2010 (see OMB, 2009). Nevertheless, we offer a robustness exploration in Chapter Five in which we present the implications of using other real interest rates.

Of course, Equations 2.1 and 2.2 are not directly comparable assuming $Y_N > Y_R$. When Y_R years have passed, the SLEPed aircraft must be replaced or SLEPed again, while the was-new aircraft will have $Y_N - Y_R$ additional years of operation remaining before a comparable decision must be made.

We do not know what the Navy's choices will be Y_R (or Y_N) years hence. Perhaps better options will be available (e.g., less-expensive or more-capable aircraft that are not currently available). But the converse could be true. Or, in between the optimistic and pessimistic cases, perhaps a new aircraft with cash flows like those of today's new aircraft will be available Y_R (and Y_N) years from now. Under this middle assumption, we can assume that the new aircraft will be replaced by its clone ad infinitum so we can translate the new aircraft's cash flows into an annuity, e.g., find the value of x_N such that

$$\sum_{t=1}^{Y_N} \frac{x_N}{(1+i)^t} = \sum_{t=0}^{Y_N} \frac{c_N(t)}{(1+i)^t}.$$

x_N would be the annual payment associated with the new aircraft being replaced by its clone indefinitely.

The SLEPed aircraft case is more complicated because it would involve Y_R years of post-SLEP service followed by either another SLEP or replacement by the new aircraft. If we define x_R such that

$$\sum_{t=1}^{Y_R} \frac{x_R}{(1+i)^t} = \sum_{t=0}^{Y_R} \frac{c_R(t)}{(1+i)^t},$$

x_R would assume that the aircraft could be repeatedly SLEPed for the same cost. This is an unrealistic assumption because eventually SLEPs would no longer be feasible and desirable. However, the inequality $x_R < x_N$ remains useful for decisionmaking about the current, proposed SLEP. If $x_R < x_N$, the annualized cost of the Y_R-year SLEP would be less than the annualized cost of the new aircraft. When Y_R years have passed, the Navy might then decide to buy the new aircraft, implying an annualized cost of x_N thereafter. Thus, the SLEP reduces Navy expenditures if $x_R < x_N$, even if the aircraft is SLEPed only once. We refer to the "cost-minimizing" choice between a Y_R-year SLEP and a new aircraft to be the choice that provides the lower of x_R and x_N, where the x's are each alternative's equivalent annuity value. Appendix B explains the equivalence of our annuitization approach with OMB's prescribed net present value calculation.

Unfortunately, as we discuss next, choosing the cost-minimizing option between a SLEP and a new aircraft ignores aircraft availability issues. There are considerable challenges in addressing availability issues, challenges that we can only partially address.

CHAPTER THREE

Consideration of Aircraft Availability Trends

As aircraft age, we expect them to cost more to maintain. See, for instance, the aging aircraft literature review provided by Dixon (2006). Also, we expect aging aircraft to be less often available for operation, due to more-frequent breakdowns and more-complex repairs. Keating and Dixon (2003), for instance, discusses downward trends in KC-135 availability, while Keating et al. (2005) shows downward drift in C-5A availability.

Unfortunately, use of the cost-minimization metric of comparing x_R and x_N, as discussed in the previous chapter, does not consider aircraft availability issues. If we plausibly assume that a new aircraft will generally have greater availability levels than a SLEPed aircraft, the cost-minimization methodology will tend to undervalue new aircraft and overvalue aircraft SLEPs.

In Keating and Dixon (2003), we proposed a way to consider aircraft availability. We defined the average cost per available year of an option (e.g., new versus SLEPed aircraft) to be

$$\frac{\sum_{t=0}^{Y} \frac{c(t)}{(1+i)^t}}{\sum_{t=1}^{Y} \frac{a(t)}{(1+i)^t}},$$

where $a(t)$ is the option's availability level in year t. The preferred option would be the choice with the lower average cost per available year.

The average-cost-per-available-year metric will yield the same findings as the cost-minimization algorithm if aircraft availability is constant over time. Suppose, for instance, that both new and SLEPed aircraft availability is constant at some level L with $0 < L \leq 1$. Then the average cost per available year in the new-aircraft case would be

$$\frac{\sum_{t=0}^{Y_N} \frac{c_N(t)}{(1+i)^t}}{L \times \sum_{t=1}^{Y_N} \frac{1}{(1+i)^t}} = \frac{x_N}{L},$$

while the average cost per available year in the SLEP case would be

$$\frac{\sum_{t=0}^{Y_R} \frac{c_R(t)}{(1+i)^t}}{L \times \sum_{t=1}^{Y_R} \frac{1}{(1+i)^t}} = \frac{x_R}{L}.$$

The average-cost-per-available-year methodology is not the same as cost minimization, however, if availability varies over time and across options. If we think that new aircraft will generally have greater availability levels than SLEPed aircraft, the average-cost-per-available-year metric will tend to be more favorable for a new aircraft than for a SLEPed aircraft.

While the average-cost-per-available-year metric is tractable and intuitive in how it adjusts for changing availability, it is not a standard, agreed-upon metric. One must accept a priori that average cost per available year is the right metric to minimize.

A different methodology would be to assume that Navy valuation of an aircraft increases linearly in $a(t)$.

If k_N denotes the multiplicative coefficient the Navy attaches to the new aircraft's availability level, it follows that the discounted net benefit of a new aircraft is

$$\sum_{t=0}^{Y_N} \frac{k_N \times a_N(t) - c_N(t)}{(1+i)^t}. \tag{3.1}$$

The formula for net benefits of a SLEPed aircraft is similar:[1]

$$\sum_{t=0}^{Y_R} \frac{k_R \times a_R(t) - c_R(t)}{(1+i)^t}. \tag{3.2}$$

As in Chapter Two, we would want to translate Equations 3.1 and 3.2 into equivalent annuities: Find the value of z_N such that

$$\sum_{t=1}^{Y_N} \frac{z_N}{(1+i)^t} = \sum_{t=0}^{Y_N} \frac{k_N \times a_N(t) - c_N(t)}{(1+i)^t}$$

and compare it to the value of z_R such that

$$\sum_{t=1}^{Y_R} \frac{z_R}{(1+i)^t} = \sum_{t=0}^{Y_R} \frac{k_R \times a_R(t) - c_R(t)}{(1+i)^t}.$$

SLEPing the aircraft is preferable if $z_R > z_N$.[2]

---

[1] k_N and k_R could, but need not, be equal. If $k_N > k_R$, the Navy would attach greater value to the new aircraft holding constant the two aircrafts' availability levels. We might typically expect new aircraft availability to exceed SLEPed aircraft availability, however.

[2] Note that, in our cost formulation in Chapter Two, one chooses the smaller of x_N and x_R. By contrast, in this formulation, we are analyzing net benefits, so one wishes to choose the greater of z_N and z_R. Of course, if one sets the k values in

If the new aircraft is sufficiently highly valued relative to the old aircraft at a constant availability level $\left(k_N \gg k_R\right)$, the outcome is clear. Implicitly, the calculations we are undertaking are sensible only if k_N and k_R are relatively close to one another in magnitude. If $k = k_N = k_R$, the new and SLEPed aircraft would be perfect substitutes if they had equal availability levels.

While we cannot know for certain what the values of these parameters are, there are bounds that can be imposed on k_N. If we assume that the new aircraft will eventually be purchased (either before or after the SLEP), it follows that

$$\sum_{t=0}^{Y_N} \frac{k_N \times a_N(t) - c_N(t)}{(1+i)^t} > 0,$$

so

$$k_N > \frac{\sum_{t=0}^{Y_N} \frac{c_N(t)}{(1+i)^t}}{\sum_{t=0}^{Y_N} \frac{a_N(t)}{(1+i)^t}}. \tag{3.3}$$

Inequality 3.3 says that k_N must be greater than the new aircraft's average cost per available year.

While the new aircraft's average cost per available year forms a lower bound on the value of k_N, there is additionally an upper bound on its value. We assume that it is appropriate to operate the new aircraft for Y_N years rather than replacing it sooner and, by assumption, replacing it with its clone. (We assume that an aircraft could be retired before Y_N years of service, if so desired, but could not be operated beyond Y_N years without a SLEP. So Y_N is a maximum non-SLEP lifetime, but not a minimum non-SLEP lifetime.)

In order for aircraft operation through and including year Y_N to be appropriate, it must be that $k_N \times a_N(Y_N) - c_N(Y_N) > z_N$, i.e., the owner is better off operating the aircraft in year Y_N, the planned last year, rather than replacing it sooner. Therefore, one must have

$$k_N \times a_N(Y_N) - c_N(Y_N) > \frac{\sum_{t=0}^{Y_N} \frac{k_N \times a_N(t) - c_N(t)}{(1+i)^t}}{\sum_{t=1}^{Y_N} \frac{1}{(1+i)^t}}.$$

Equations 3.1 and 3.2 equal to 0, one finds $x_N = -z_N$ and $x_R = -z_R$.

This inequality then reduces to

$$\frac{\left(\frac{\sum_{t=0}^{Y_N}\frac{c_N(t)}{(1+i)^t}}{\sum_{t=1}^{Y_N}\frac{1}{(1+i)^t}}-c_N(Y_N)\right)}{\left(\frac{\sum_{t=0}^{Y_N}\frac{a_N(t)}{(1+i)^t}}{\sum_{t=1}^{Y_N}\frac{1}{(1+i)^t}}-a_N(Y_N)\right)}>k_N. \tag{3.4}$$

The ratio in the numerator of Inequality 3.4 is the new-aircraft average cost per year. Likewise, in the denominator, the ratio is the aircraft's average availability level per year. If we assume that availability trends down with age, year Y_N will have below-average availability, so the denominator is a positive number. The numerator must also be positive, i.e., year Y_N's cost must be below the life-cycle average cost (that includes year 0 procurement costs).

As we illustrate in Chapter Five, one can gain some insight implementing this z_R versus z_N methodology, most especially if one is willing to assume that $k = k_R = k_N$. As is the case with the average-cost-per-available-year metric, use of the $k \times a(t) - c(t)$ objective function makes new aircraft more desirable than under Chapter Two's cost minimization.

There are, however, concerns that remain with this new methodology. Even if one assumes that $k = k_R = k_N$, one still gets a range of cutoffs for SLEP desirability with the different values coming from the minimum possible k value (Inequality 3.3) and the maximum possible k value (Inequality 3.4). How should one interpret, for instance, a SLEP being worthwhile if k is at its minimum value but not if k is at its maximum value? There is no clear way to assess the Navy's true value of k.

An additional problem with this methodology is that it assumes that the Navy's objective function is linear in availability. Suppose, alternatively, that the Navy had diminishing marginal utility in aircraft availability, e.g.,

$$p \times \sqrt{a(t)} - c(t).$$

We would then have different constraints on the values of p_N.

Next we present F/A-18E/F data and, later, in Chapter Five, discuss what our different methodologies suggest about the desirability of F/A-18E/F SLEPs.

CHAPTER FOUR

F/A-18E/F Context

F/A-18s are carrier-capable fighter attack aircraft manufactured by McDonnell Douglas, now Boeing.[1] There have been three generations of F/A-18s: the A/B version that began operation in October 1983, the C/D version that began operation in September 1987, and the E/F version that achieved initial operational capability in September 2001. The A, C, and E versions are one-seat aircraft; the B, D, and F models are two-seat aircraft. (The second seat can be occupied by a training officer or by a combat officer performing nonpilot functions, such as tactical or forward air control or controlling the aircraft's weapons.) The A–Ds are referred to as Hornets, while the Es and Fs are referred to as Super Hornets.

The Super Hornets will be due for a SLEP or for retirement sometime late in the 2010 decade or early in the 2020 decade. It is this SLEP decision about which the Navy requested insight.

The models presented in Chapters Two and Three are informationally demanding as enumerated in Table 4.1. One needs an estimate of the cost of the SLEP as well as the additional years of operation it allows. One also wants to know the life-cycle costs (procurement and year-to-year maintenance and modification costs[2]) associated with the replacement aircraft alternative to doing an E/F SLEP. If aircraft availability trends are to be considered, one would also need to estimate the levels of aircraft availability associated with both a post-SLEP F/A-18E/F and its prospective replacement.

Unfortunately, an F/A-18E/F service life assessment program only recently commenced. (Such an assessment is a necessary prerequisite to doing a SLEP.) So, many of the parameters we need are not known and can only be estimated with a high degree of imprecision. Our philosophy is to start with "best guesses" as to appropriate parameter values then to undertake extensive robustness analysis to identify the most-decisive parameters. There is little value in worrying about an imprecisely estimated parameter if reasonable perturbations in its value do not appear to alter decisions.

We assume that the alternative to undertaking E/F SLEPs is to acquire the F-35C carrier variant of the Joint Strike Fighter (JSF). An alternative that could be considered for future analysis would be buying new F/A-18E/Fs as an alternative to E/F SLEPs. O'Rourke (2009) notes Navy plans for the final procurement of F/A-18E/Fs in fiscal years (FYs) 2010–2012, so there would be a sizable gap between the Navy's currently planned end of E/F procurement and late in this decade, when current E/Fs will begin to require either SLEPs or replacement.

[1] This paragraph's information is taken from U.S. Navy (2009).

[2] We focus only on maintenance and modification costs here. We assume that other operating and support (O&S) costs, e.g., mission personnel costs, do not vary with aircraft age nor between the SLEPed aircraft and the new aircraft.

Table 4.1
Parameters to Be Estimated

Aircraft	Parameter	Description
SLEP	Y_R	Years of additional service post-SLEP
	$c_R(t)$	Annual constant-dollar costs of post-SLEP aircraft
	$c_R(0)$	Per-aircraft cost of SLEP
	$c_R(1 \ldots Y_R)$	Annual maintenance and modification costs of SLEPed aircraft
	$a_R(t)$	Annual availability levels of SLEPed aircraft
New	Y_N	Years of service provided by new aircraft
	$c_N(t)$	Annual constant-dollar costs of new aircraft
	$c_N(0)$	Per-aircraft procurement cost
	$c_N(1 \ldots Y_N)$	Annual maintenance and modification costs of new aircraft
	$a_N(t)$	Annual availability levels of new aircraft

But this option is not completely implausible, since E/F foreign-military sales will continue for a period past the cessation of U.S. Navy procurement of this aircraft. Scully (2010) notes recent discussions about the Navy possibly buying more F/A-18E/Fs.

If E/F SLEPs are undertaken, it is possible that the Navy could skip JSFs (at least as F/A-18E/F replacements) and jump to the so-called next-generation F/A-XX, a strictly conceptual aircraft that, for instance, may or may not be manned. However, procuring an F/A-XX is not a realistic option in time to supplant the E/F SLEP due toward the end of this decade. It could be an option, however, to replace a once-SLEPed E/F a decade or so later. For modeling purposes, we assume that the Navy will eventually replace the F/A-18E/F fleet with JSFs. The issue of interest is whether to do so before or after an E/F SLEP.

Best Estimates of Model Parameters

Our task then becomes to estimate values of the parameters in Table 4.1. Unfortunately, our discussions with Navy experts found that many of the parameters we need have not been estimated with precision.

We were told that the E/F SLEP under consideration might add ten years to the aircraft's life, so $Y_R = 10$. However, considerable uncertainty was expressed about this estimate, so we will assess robustness of findings to changes in this parameter estimate.

There is no final estimate of the cost of an E/F SLEP, but we were told that F/A-18C/D SLEPs, which are to commence in the next year or two, are estimated to cost \$26 million (FY 2009 dollars) per aircraft.

There are conflicting theories as to whether an E/F SLEP would cost more or less than a C/D SLEP. The E/F is a larger aircraft with approximately a one-third greater flyaway cost[3] than the C/D variant, suggesting that its SLEP might cost more than a C/D SLEP.[4] On the other hand, Patuxent River personnel suggested that the E/F is a better-maintained aircraft without the center barrel problems that have afflicted the C/D. Balancing these arguments, we therefore assume that an E/F SLEP will cost the same as a C/D SLEP, so $c_R(0) = \$26M$, though, as with Y_R, this is a high-uncertainty parameter calling for extensive robustness analysis.

Since E/Fs are currently relatively new aircraft, it is hard to project their annual maintenance and modification costs as they near the SLEP (not to mention after undergoing one). Navy Visibility and Management of Operating and Support Cost (VAMOSC) data show E/F maintenance and modification expenditures per aircraft of about $2.5 million in FY 2008. But the E/F fleet was relatively young in 2008, with an average aircraft age of about five years. What will happen to maintenance and modification costs as this aircraft ages?

VAMOSC data on the C/Ds, shown in Figure 4.1, present cause for concern. In this figure, we show constant-dollar F/A-18C and D maintenance and modification expenditures

Figure 4.1
F/A-18C/D Maintenance and Modification Expenditures per Aircraft, 1989–2008

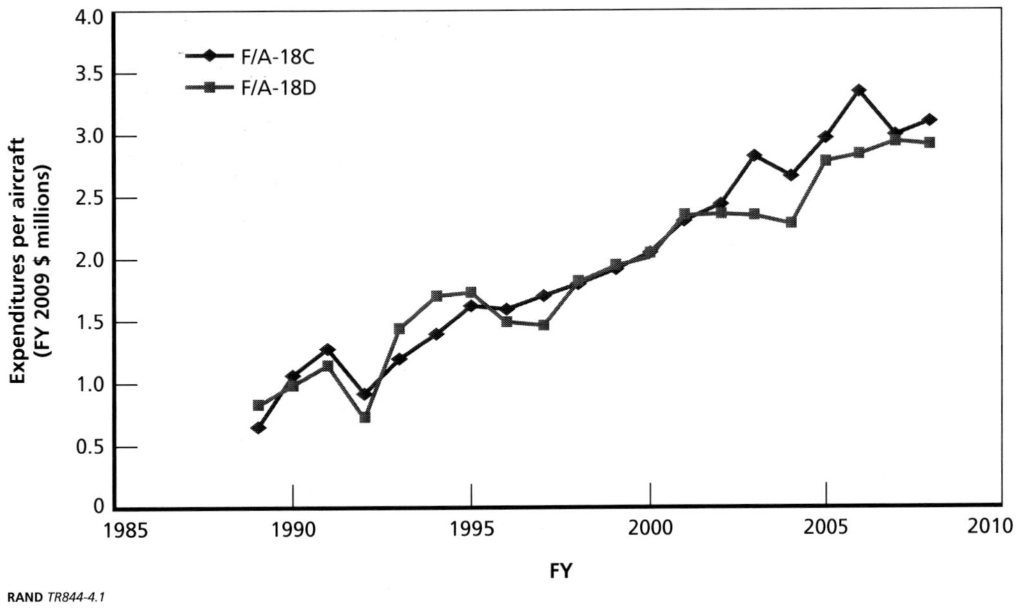

RAND *TR844-4.1*

[3] The Defense Acquisition University (undated) defines *flyaway cost* as the cost of procuring prime mission equipment. Flyaway cost does not, however, include the costs of support items and initial spares. Those additional costs are added to flyaway costs to form procurement unit cost. We prefer to use procurement unit cost in our analysis because those support items and initial spares are typically purchased when additional aircraft are purchased, so they should be included if we are estimating an aircraft's marginal cost.

[4] Pyles (2003) uses aircraft flyaway cost as an independent variable in a number of aircraft maintenance workload and material-consumption regressions. In regressions with on-equipment workload, base periodic-inspection workload, special inspection workload, material consumption, contractor logistics support costs, and time-change technical order workload as dependent variables, he finds flyaway cost to have positive and statistically significant coefficients, i.e., workload and therefore expenditures are greater in aircraft with greater flyaway costs.

per aircraft between 1989 and 2008. VAMOSC data commence in 1989, so we do not observe the initial years of C/D operation.

Taken at face value, Figure 4.1 presents a fairly alarming pattern of constant-dollar increases. The D's annual maintenance and modification expenditures have grown at nearly a 7-percent annual rate since 1989; the C's annual maintenance and modification expenditure growth rate has exceeded 8 percent. The rates of maintenance and modification expenditure per aircraft increases shown in Figure 4.1 are so large that it is unlikely that they solely represent aging effects. For example, the Congressional Budget Office (Kiley and Skeen, 2001) found that "studies typically have found that the costs of operating and maintaining aircraft increase by 1 percent to 3 percent with every additional year of age, after adjusting for inflation" (pp. 21–22).

So we believe that actual "aging effects" are not like those depicted in Figure 4.1, but, at the same time, it is reasonable to suspect that some (perhaps small) component of the observed pattern of real expenditure growth is age related.

We decided to parameterize E/F maintenance and modification expenditures per aircraft as being constant at $2.5 million per year during the first five years of aircraft life followed by a 3-percent real rate of increase annually starting in the sixth year of operation (so expenditures per aircraft would approach $4 million in the aircraft's 20th year of operation). Figure 4.2 depicts this assumed real expenditure per aircraft pattern.

We have also superimposed VAMOSC data on actual F/A-18E and F/A-18F maintenance and modification expenditures per aircraft for 2004–2008. VAMOSC shows a sharper increase in the early years of aircraft operation than we are assuming. As noted earlier, we do not think that all of the maintenance and modification expenditure per aircraft increase observed in VAMOSC is caused by age effects.

Figure 4.2
Assumed F/A-18E/F and Joint Strike Fighter Maintenance and Modification Expenditures per Aircraft, as a Function of Aircraft Age

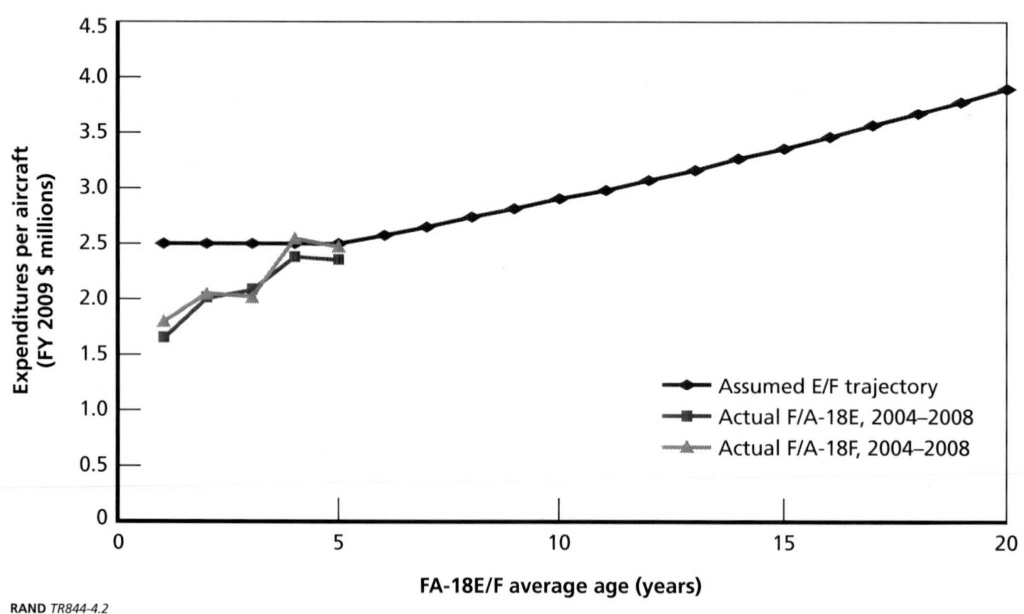

RAND *TR844-4.2*

A further challenge is to estimate how maintenance costs (and availability levels—the challenge is symmetric) might change after the aircraft's SLEP.

We have identified three contrasting theories as to the effect of SLEPs on aircraft maintenance costs and availability levels. The most pessimistic is that a SLEP has no favorable effect on either data series, e.g., the adverse pattern assumed in Figure 4.2 will continue past age 20. This theory seems unduly pessimistic, however, since the SLEP will presumably resolve a number of nagging maintenance issues, leaving the post-SLEP aircraft in at least somewhat better condition.

A diametrically opposite theory would be that the SLEP "zero hours" the aircraft, i.e., it would then have the annual maintenance and modification costs associated with a new aircraft. This theory strikes us as being illogically optimistic if $Y_R < Y_N$. If we assume that the aircraft will operate only Y_R years after the SLEP, it seems untenable to believe that it would have the maintenance costs and availability level associated with a new aircraft with Y_N years of remaining life.

Our middle-ground preferred assumption is that the SLEP "pulls back" annual expenditures per aircraft in accord with the number of extra years of service provided by the SLEP. For instance, a ten-year SLEP after 20 years of operation results in the aircraft having the annual maintenance and modification expenditures of an 11-year-old aircraft (actual age of 22 minus one year spent in the SLEP minus ten years for the ten extra years of service provided by the SLEP). Therefore, we assume that $c_R(1...10) = c(11...20)$, where $c(11...20)$ are our assumed annual E/F maintenance and modification costs per aircraft between ages 11 and 20.

We were unable to find evidence either supporting or refuting this "pullback" assumption. We believe that it is a more logical assumption than either continuing cost trends as if the SLEP did not occur or "zero-houring" the aircraft. Perhaps future analysis could examine the cost and availability implications of the soon-to-commence C/D SLEP.

If we are to consider aircraft availability issues in our analysis using either of Chapter Three's methodologies, we must project the rates at which SLEPed E/Fs and JSFs will be available for and capable of performing missions.

There are two requirements for an aircraft to be useful to the Navy. The first is that it be operator possessed, i.e., not tied up in the depot-level maintenance system. The second requirement is that the operator-possessed aircraft be deemed to be mission capable. The second criterion is embedded in the well-known mission-capability (MC) rate. But MC rates are generally computed conditional on being operator possessed. A fleet of aircraft could have a very high MC rate but nevertheless be providing poor performance to the Navy if a high percentage of aircraft was not operator possessed.

Figure 4.3 shows the F/A-18C/D's quarterly equipment-in-service (EIS) rate, i.e., operator possession rate, from the first quarter of FY 1990 through the third quarter of FY 2009, with the corresponding average ages of the C/D fleet on the horizontal axis. Figure 4.3 was derived by combining Aircraft Inventory Readiness and Reporting System (AIRRS) data on the number of F/A-18C/Ds owned by the Navy, AIRRS data on quarterly average fleet ages, and Decision Knowledge Programming for Logistics Analysis and Technical Evaluation (DECKPLATE) data on the number of monthly (aggregated into quarterly to match AIRRS) EIS hours in the F/A-18C/D fleet. For example, in the first quarter of FY 1990 (October–December 1989), DECKPLATE shows 274,954 F/A-18C EIS hours, which would correspond to about 124.5 aircraft. But AIRRS shows that the Navy owned 146 F/A-18Cs as of December 1989, so their implied EIS rate for the first quarter of FY 1990 was about 85 percent. The

Figure 4.3
F/A-18C/D Equipment-in-Service Rate, as a Function of Average Fleet Age

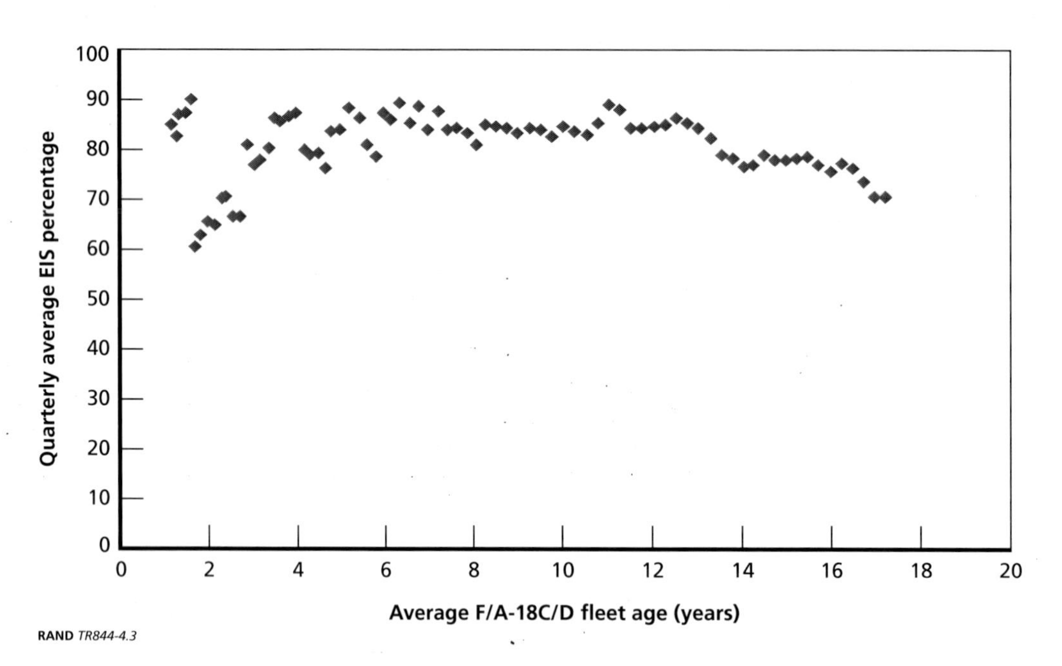

aircraft that were not registering EIS hours were in depot-level maintenance, we were told. AIRRS also provides the average fleet age by quarter, forming the horizontal axis of Figure 4.3.

EIS rates were volatile early in the C/D fleet's life and stabilized between roughly age 5 and age 12 but have declined considerably in recent years, i.e., a greater percentage of the fleet has been tied up in depot maintenance.

Figure 4.4 shows the F/A-18C/D's quarterly MC rates over the same time period. DECKPLATE records MC rates; average fleet ages come from AIRRS. MC rates are conditional on fleet possession, i.e., they cover only those aircraft that are not currently in depot maintenance.

F/A-18C/D MC rates have been more stable than EIS rates, although, again, downward drift is evident.

In Keating et al. (2005), we defined a fleet's composite availability rate to be the product of the EIS and MC rates, i.e., the fraction of the total fleet held by operators and mission capable. Figure 4.5 shows the F/A-18C/D's composite availability rates over time. Figure 4.5 simply multiplies the rates shown in Figures 4.3 and 4.4.

We estimated a regression of the natural log of C/D composite availability on aircraft age covering ages 5–17. The age coefficient in that regression was about –0.023, i.e., there is a bit over a 2-percent annual downward drift in the C/D's composite availability rate.

Figure 4.5 also shows the F/A-18E/F's composite availability rates as of the same ages. Of course, we have many fewer E/F data.

After a difficult start, E/F composite availability in recent years (as the average E/F fleet age has neared five years) has roughly equaled C/D composite availability at similar average fleet ages. Our baseline assumption, therefore, is that E/F composite availability will follow the pattern observed in the C/D fleet.

As with maintenance costs, we assume that a ten-year SLEP rolls the aircraft's availability up to the level it was at 11 years prior. Therefore, we assume that $a_R(1...10) = a(11...20)$.

Figure 4.4
F/A-18C/D Mission-Capability Rate, as a Function of Average Fleet Age

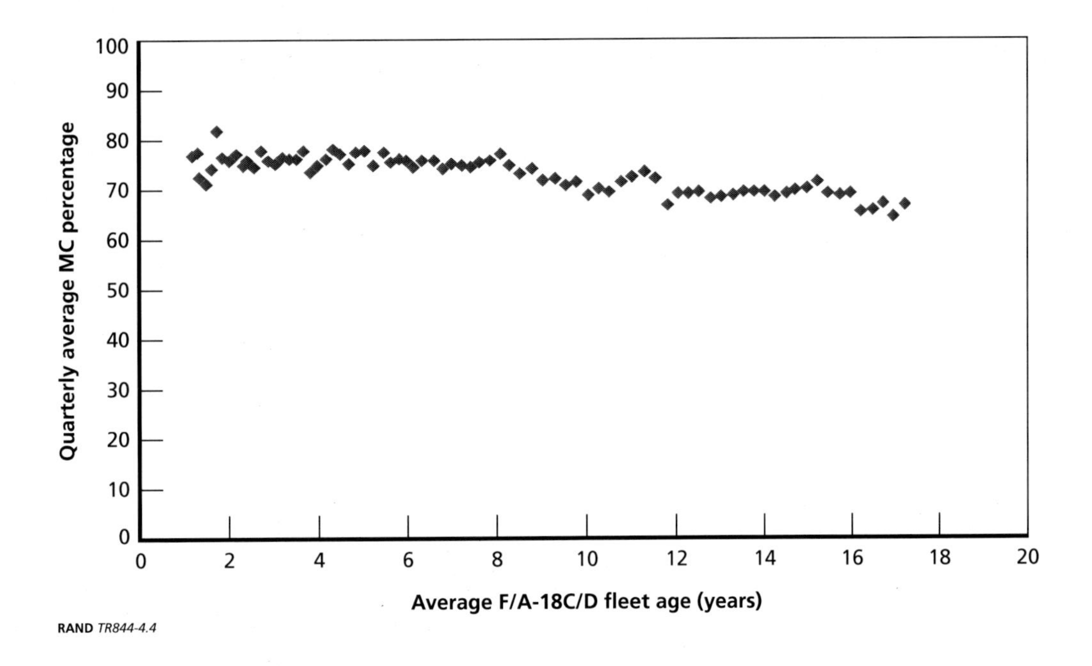

RAND *TR844-4.4*

Figure 4.5
F/A-18C/D and E/F Composite Availability, as a Function of Average Fleet Age

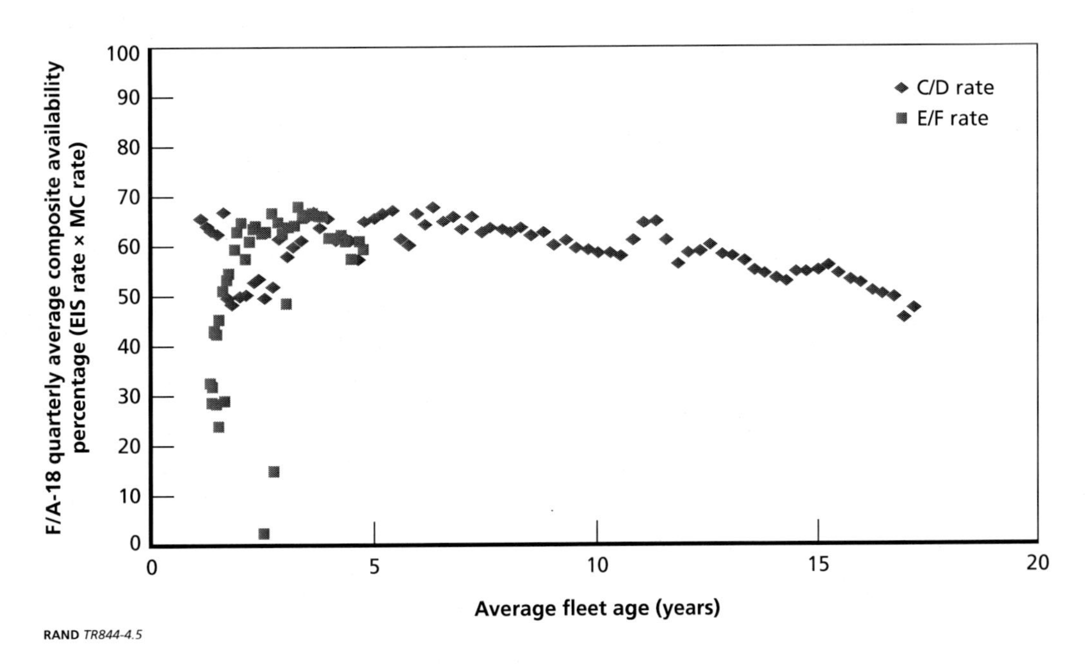

RAND *TR844-4.5*

We must also make assumptions about Table 4.1's JSF parameters, the new-aircraft alternative in this analysis.

The Navy told the RAND research team to assume that a JSF will have a 30-year lifetime, so $Y_N = 30$.

We assume that a JSF's procurement unit cost will be \$80 million, so $c_N(0) = \$80M$. The December 2007 Selective Acquisition Report indicates a JSF average procurement unit cost of about \$69 million in budget-year 2002 dollars or roughly \$83 million in FY 2009 dollars using a 20-percent increase in the Bureau of Economic Analysis' gross domestic product (GDP) deflator between FY 2002 and FY 2009 (see Bureau of Economic Analysis, 2009). There have been concerns expressed that JSF costs have escalated since December 2007. See, for instance, Drew (2009). Our \$80 million baseline JSF procurement unit cost might, unfortunately, be quite optimistic.

We assume that JSF maintenance and modification costs will be the same as an E/F's at the same age. We assume that $c_N(1...20) = c(1...20)$, then further assume that $c_N(21...30)$ continues Figure 4.2's 3-percent real growth. The assumption that JSF maintenance and modification costs will follow Figure 4.2's trajectory might again be optimistic, i.e., pro-JSF. Sherman (2010) notes concerns that JSF O&S costs might exceed F/A-18 support costs.

We assume that JSF composite availability will equal assumed E/F availability through age 20 and then stabilize. If one lets composite availability continue to decline at 2.3 percent per annum past age 20, one would estimate a 37.9-percent composite availability rate in year 30. Such a value, however, would imply that Inequality 3.4's maximum value of k_N is less than Inequality 3.3's minimum value of k_N, given our other parameter estimates. Having it be optimal to operate the JSF for 30 years is incompatible with composite availability continuing to decline on the trajectory estimated using C/D ages 5–17 composite availability data, we find.

Chapter Summary

Summarizing the information presented in this chapter, we present our baseline parameters for a ten-year E/F SLEP versus a new JSF.

Table 4.2 presents our assumed \$26 million ten-year E/F cash flow and composite availability levels.

The alternative to doing an E/F SLEP would be to acquire a new JSF. Table 4.3 presents our assumed JSF life-cycle cash flow and composite availability levels.

Note that the assumed 30-year JSF life span does not include a requirement to have a SLEP at age 21. As a result of not having that SLEP, the age 22–30 JSF has considerably greater maintenance and modification costs than a SLEPed E/F of the same age, e.g., \$2.99 million for a 22-year-old SLEPed E/F versus \$4.13 million for a 22-year-old non-SLEPed JSF. However, we assume that composite availability stabilizes at 47.7 percent after year 20.

Using the parameters described in this chapter, Chapter Five presents different methodologies' assessments of the desirability of E/F SLEPs.

Table 4.2
Assumed Ten-Year E/F Service Life Extension Program Parameters

Year	E/F Age (years)	Maintenance and Modification Costs (FY 2009 $ millions)	Composite Availability Percentage	Comment
0	21	26	0	Aircraft is out of service for one year receiving the SLEP.
1	22	2.99	58.6	These are the assumed costs and availability levels for aircraft 11 years younger.
2	23	3.07	57.3	
3	24	3.17	56.0	
4	25	3.26	54.7	
5	26	3.36	53.5	
6	27	3.46	52.3	
7	28	3.56	51.1	
8	29	3.67	49.9	
9	30	3.78	48.8	
10	31	3.89	47.7	

Table 4.3
Assumed 30-Year Joint Strike Fighter Parameters

Year	JSF Age (years)	Procurement, Maintenance, and Modification Costs (FY 2009 $ millions)	Composite Availability Percentage
0	0	80	0
1	1	2.50[a]	52.5[b]
2	2	2.50[a]	62.8[b]
3	3	2.50[a]	65.2[b]
4	4	2.50[a]	60.1[b]
5	5	2.50[a]	67.2[c]
6	6	2.58[a]	65.7[c]
7	7	2.65[a]	64.2[c]
8	8	2.73[a]	62.8[c]
9	9	2.81[a]	61.3[c]
10	10	2.90[a]	60.0[c]
11	11	2.99[a]	58.6[c]
12	12	3.07[a]	57.3[c]
13	13	3.17[a]	56.0[c]

Table 4.3—Continued

Year	JSF Age (years)	Procurement, Maintenance, and Modification Costs (FY 2009 $ millions)	Composite Availability Percentage
14	14	3.26[a]	54.7[c]
15	15	3.36[a]	53.5[c]
16	16	3.46[a]	52.3[c]
17	17	3.56[a]	51.1[c]
18	18	3.67[a]	49.9[c]
19	19	3.78[a]	48.8[c]
20	20	3.89[a]	47.7[c]
21	21	4.01[d]	47.7[e]
22	22	4.13[d]	47.7[e]
23	23	4.26[d]	47.7[e]
24	24	4.38[d]	47.7[e]
25	25	4.52[d]	47.7[e]
26	26	4.65[d]	47.7[e]
27	27	4.79[d]	47.7[e]
28	28	4.93[d]	47.7[e]
29	29	5.08[d]	47.7[e]
30	30	5.23[d]	47.7[e]

[a] From Figure 4.2.

[b] Actual E/F composite availability.

[c] From C/D availability/age regression.

[d] Continued 3% real growth.

[e] Assumed static composite availability past age 20.

CHAPTER FIVE

F/A-18E/F Service Life Extension Program Desirability Analysis

Chapter Two presented our cost-minimization methodology to assess the desirability of a SLEP, while Chapter Three presented minimization of average cost per available year as well as a net benefit maximization methodology. In this chapter, we use the parameters presented in Chapter Four to assess F/A-18E/F SLEP desirability using all three methodologies.

There are three broad purposes to this chapter: assessment of F/A-18E/F SLEP desirability, illustration of how the different methodologies work, and comparison of how the methodologies affect the findings.

We start with cost minimization.

Cost-Minimization Methodology

Given our best estimates of relevant parameters, if the Navy's objective is to minimize annuitized costs, the $26 million ten-year SLEP is the preferred approach. Using the data in Table 4.2 in Chapter Four, the annualized cost (x_R) of the ten-year SLEP is about $6.4 million. The data in Table 4.3 indicate a JSF annualized cost (x_N) of about $7.3 million.

But more than "the answer," we are interested in how answers vary as parameter estimates vary. Many parameters important to decisionmaking are unknown. We therefore explore varying key parameters, such as the cost of the SLEP, the years of additional operation provided by the SLEP, and the cost of acquiring a new aircraft, to see how they alter decisionmaking.

In Figure 5.1, we present the annualized cost associated with an E/F SLEP versus a JSF as a function of the cost of a ten-year SLEP. A SLEP that adds ten years of operation to an E/F can cost up to about $33.6 million, we find, and still reduce Navy costs. We have placed the vertical axis at our baseline assumption that a ten-year SLEP would cost $26 million.

In Figure 5.2, we vary the number of extra years of operation emanating from a $26 million SLEP. We find that a $26 million SLEP reduces costs if it provides eight or more years of extra operation. We have placed the vertical axis at our baseline assumption that the SLEP will provide ten additional years of operation.

Another important parameter is the JSF's procurement unit cost. In Figure 5.3, we vary the JSF's procurement unit cost and evaluate its effect on the relative cost of an E/F SLEP. Our baseline assumption is that the JSF's procurement unit cost will be $80 million.

If the JSF has a procurement unit cost greater than $62.1 million, we find that doing a ten-year $26 million E/F SLEP would reduce Navy costs.

We also assessed the importance of the real interest rate in SLEP decisionmaking. As discussed in Chapter Two, our interpretation of OMB Circular A-94 is that Navy decision-

Figure 5.1
Estimated Cost Implications of a Service Life Extension Program That Adds Ten Years of E/F Operation

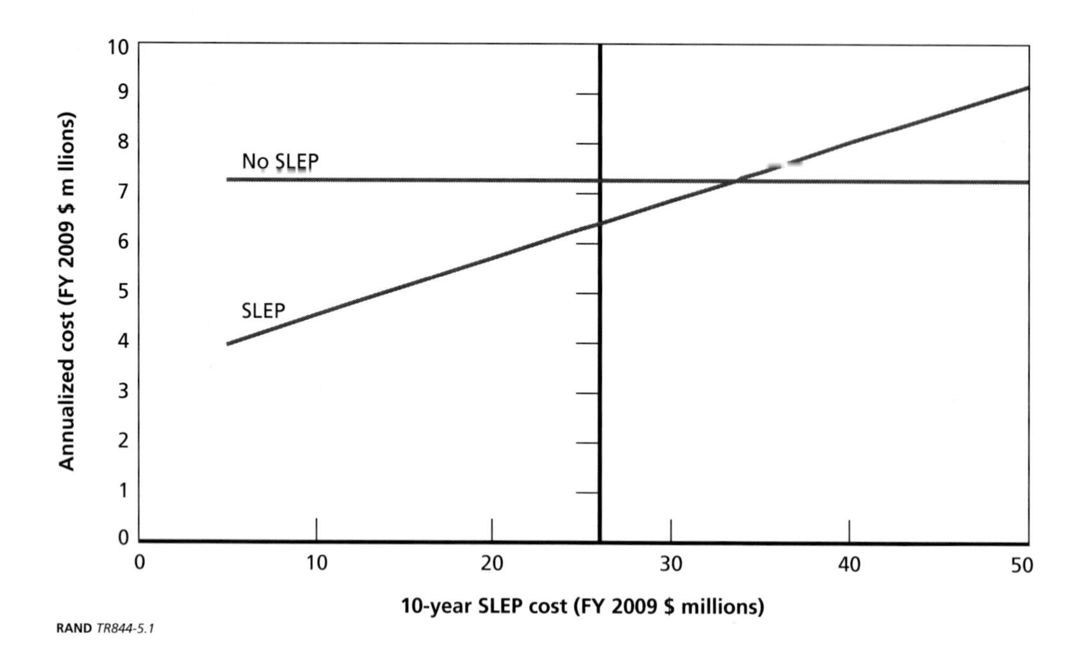

RAND *TR844-5.1*

Figure 5.2
Comparative Cost Implications of a $26 Million Service Life Extension Program, as a Function of Years of Post-SLEP Operation

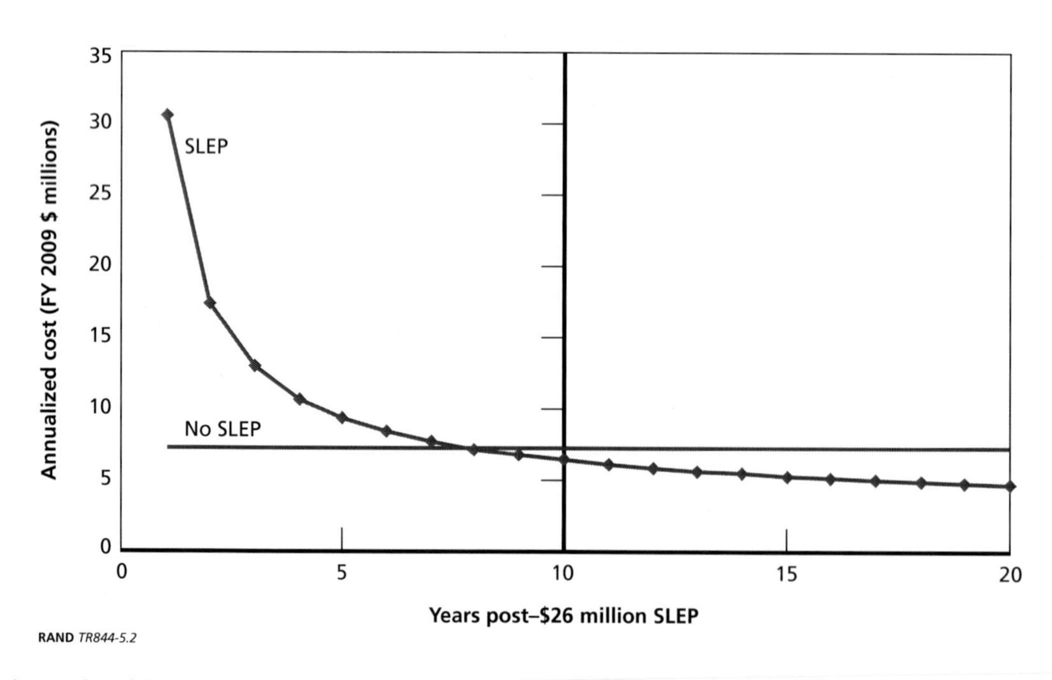

RAND *TR844-5.2*

making should use the 2010-prescribed 2.7-percent long-term real interest rate. Some observers believe that the U.S. Department of Defense's (DoD's) budget process leads decisionmakers to give near-term costs and benefits more emphasis than a 2.7-percent real interest rate would sug-

Figure 5.3
F/A-18E/F Service Life Extension Program Relative Cost, as a Function of Joint Strike Fighter Procurement Unit Cost

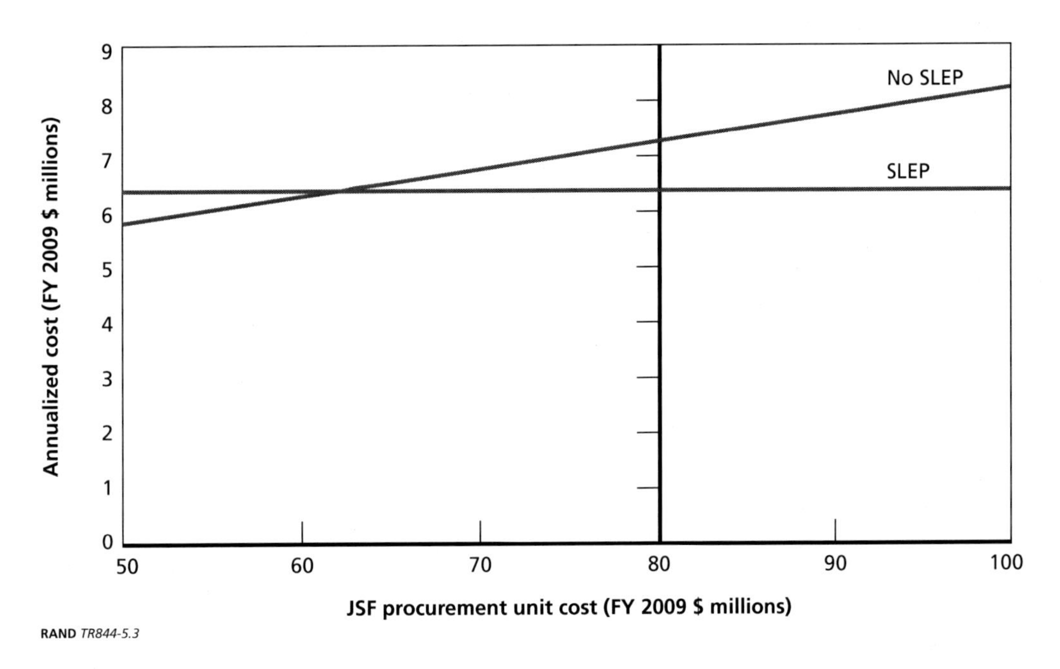

RAND *TR844-5.3*

gest. For example, a 2.7-percent real interest rate implies that a constant-dollar flow 30 years from now is worth $0.45 in today's dollars

$$\left(\frac{1}{1.027^{30}} \approx 0.45\right).$$

By contrast, a 10-percent real interest rate would imply that a dollar 30 years from now is worth less than $0.06 in today's dollars

$$\left(\frac{1}{1.1^{30}} \approx 0.057\right).$$

If DoD decisionmakers pay little attention to flows that might occur 30 years from now, we might infer that these decisionmakers have effective real interest rates greater than the prescribed 2.7 percent. As illustrated in Figure 5.4, higher real interest rates tend to favor SLEPs over aircraft replacement, which implies greater expenditures in the short term. We have placed the vertical axis at the prescribed 2.7-percent real interest rate and the horizontal axis at our baseline assumed ten-year SLEP cost of $26 million.

The intuition for Figure 5.4 is that, as the real interest rate increases, the Navy becomes less willing to pay the up-front costs of a new aircraft and therefore more willing to pay the lower cost of a SLEP instead.

We also evaluated the sensitivity of findings to the maintenance and modification expenditure growth rate discussed in Figures 4.1 and 4.2. As shown in Figure 5.5, the maximum willingness to pay for a ten-year SLEP is essentially unaffected by the assumed maintenance

Figure 5.4
Cost-Minimization Maximum Willingness to Pay for a Ten-Year E/F Service Life Extension Program, as a Function of the Real Interest Rate

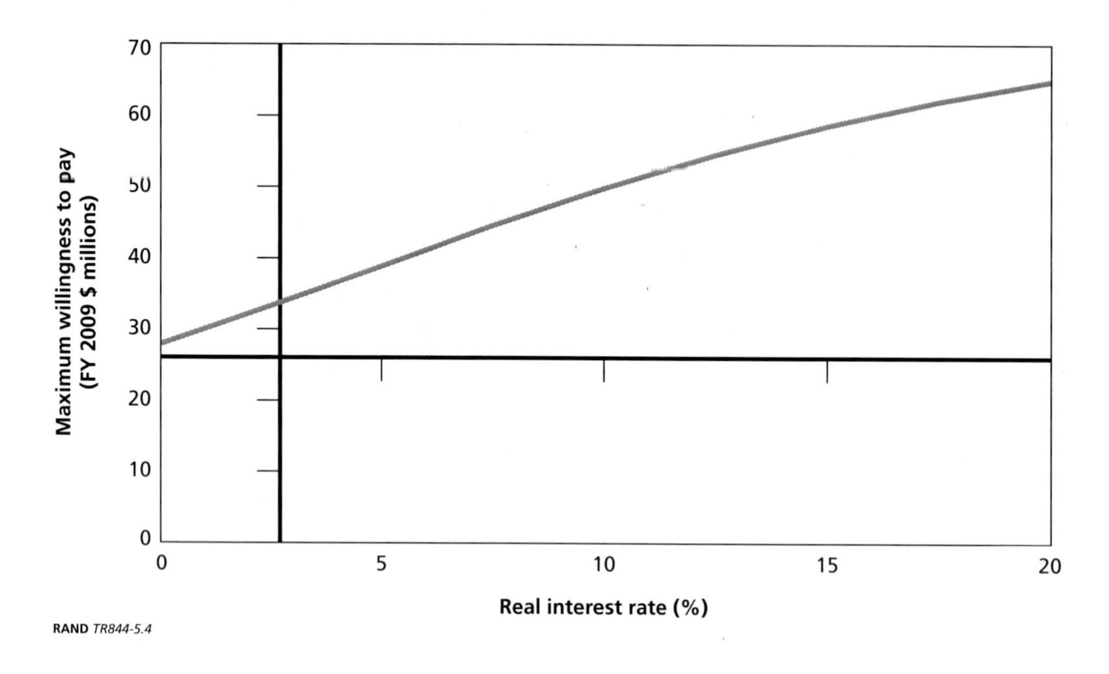

RAND *TR844-5.4*

Figure 5.5
Cost-Minimization Maximum Willingness to Pay for a Ten-Year E/F Service Life Extension Program, as a Function of the Maintenance and Modification Expenditure Growth Rate

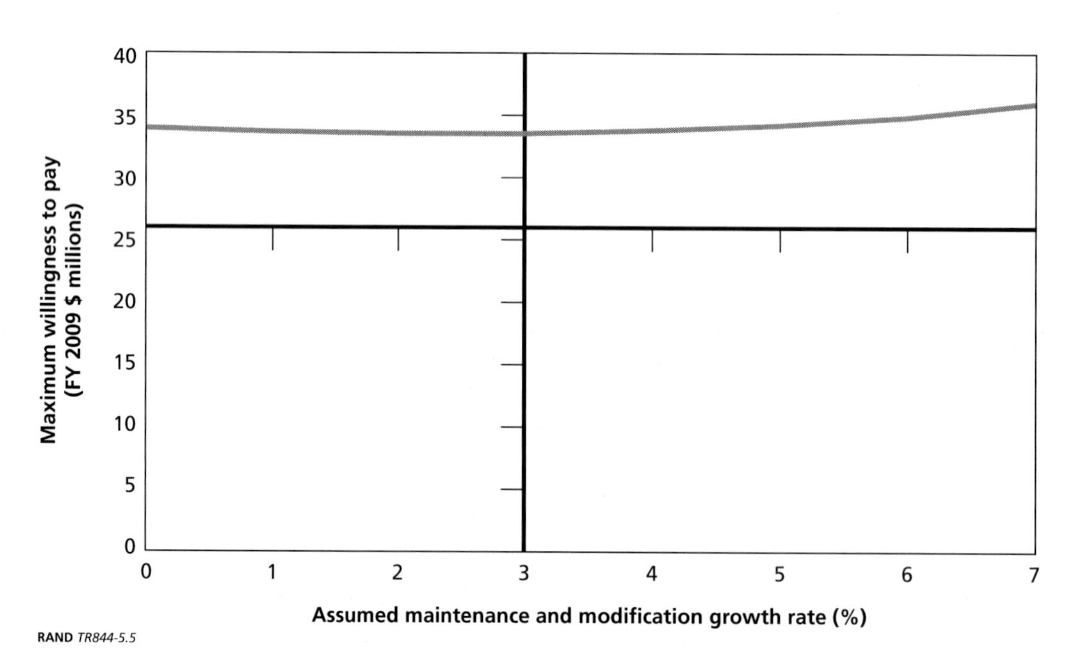

RAND *TR844-5.5*

and modification cost growth rate. The vertical axis is placed at our assumed 3-percent growth rate past age 5. The horizontal axis is placed at our assumed $26 million ten-year E/F SLEP cost.

A greater cost growth rate makes E/F post-SLEP years more expensive, but it also makes the JSF (which is assumed not to receive a SLEP) more expensive. Life-cycle costs of both systems are greater with faster-growing maintenance and modification costs, but the assumed growth rate has no important impact on the decision whether to undertake E/F SLEPs.

Minimization of Average Cost per Available Year

Given our parameter assumptions, if the Navy's objective is to minimize average cost per available year, the $26 million ten-year SLEP is again the preferred approach. Using the data in Tables 4.2 and 4.3 in Chapter Four, the average cost per available year of the ten-year SLEP is about $12.0 million, whereas the average cost per available year of a new JSF is about $13.1 million.

As above, we want to assess the sensitivity of results to different parameter estimates. In Figure 5.6, we present the average cost per available year associated with an E/F SLEP as a function of the cost of a ten-year SLEP. A SLEP that adds ten years of operation to an E/F can cost up to about $30.9 million, we find, and still reduce Navy costs per available year.

The y-axis scales of Figure 5.1 and Figure 5.6 are different. Average cost per available year is consistently greater than annualized cost because composite availability rates are around 50 percent. Figure 5.6 also has its SLEP/no SLEP crossing point somewhat to the left, i.e., at a somewhat lower SLEP cost. Since new aircraft are assumed to have generally greater availability rates than SLEPed aircraft, consideration of availability makes SLEPs less desirable, i.e., reduces the SLEP cost breakeven point, in this case from around $33.6 million to around $30.9 million.

Figure 5.6
Estimated Average-Cost-per-Available-Year Implications of a Service Life Extension Program That Adds Ten Years of E/F Operation

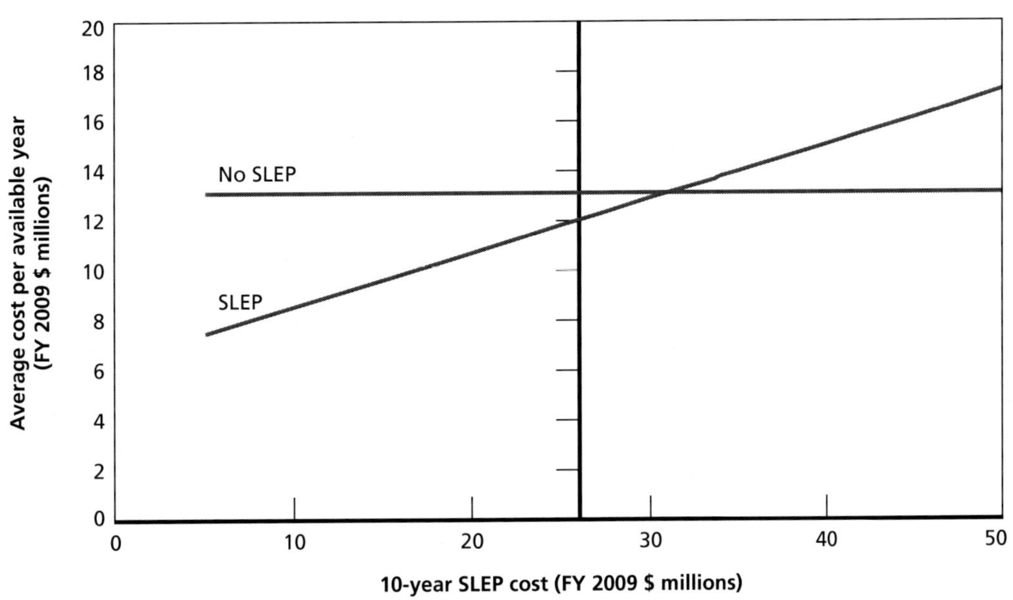

RAND *TR844-5.6*

In Figure 5.7, we vary the number of extra years of operation emanating from a $26 million SLEP. We find that a $26 million SLEP reduces average cost per available year if it provides nine or more years of operation, up one year from Figure 5.2.

Finally, Figure 5.8 assesses the importance of the JSF's procurement unit cost.

Figure 5.7
Comparative Average-Cost-per-Available-Year Implications of a $26 Million Service Life Extension Program, as a Function of Years of Post-SLEP Operation

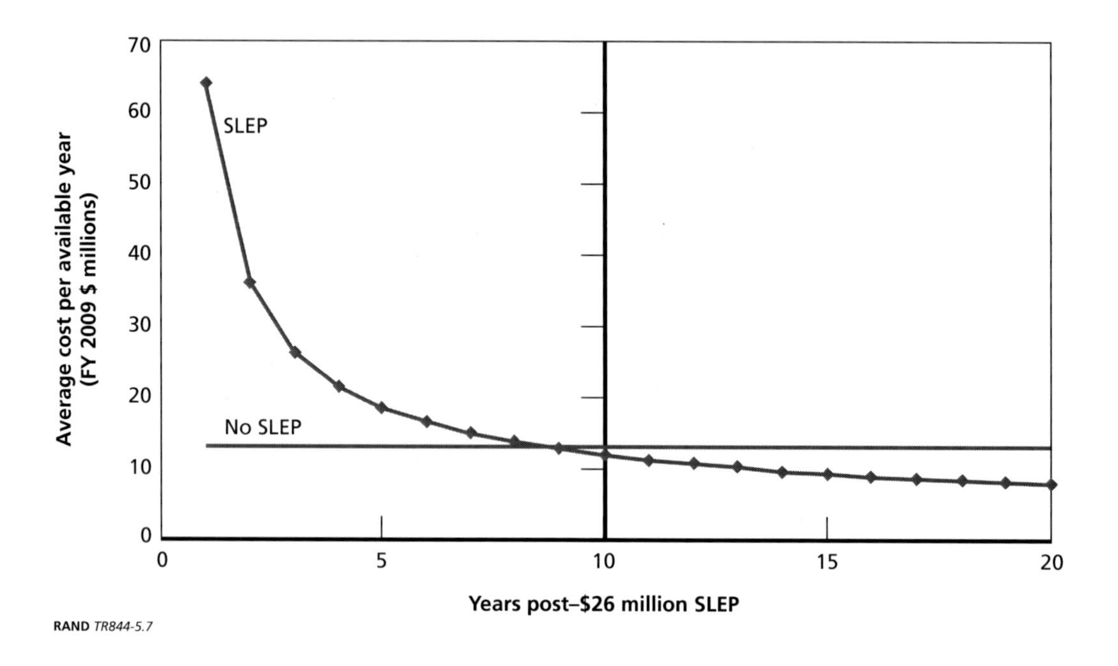

Figure 5.8
F/A-18E/F Service Life Extension Program Relative Average Cost per Available Year, as a Function of Joint Strike Fighter Procurement Unit Cost

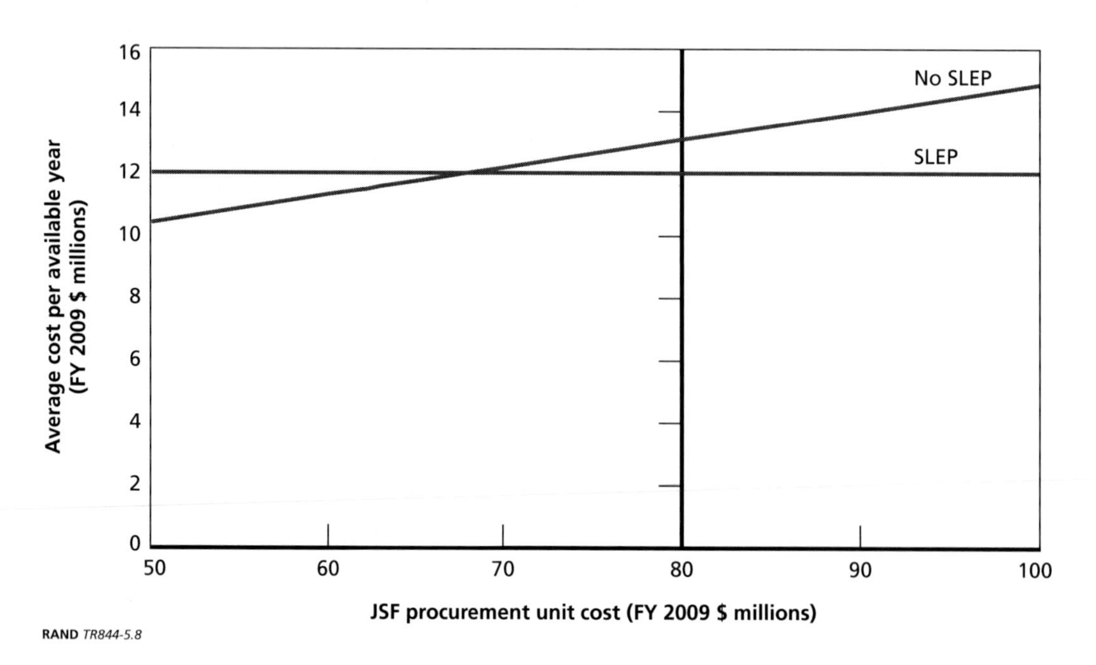

Whereas Figure 5.3 found a breakeven JSF procurement unit cost of \$62.1 million, Figure 5.8 suggests that a \$26 million ten-year E/F SLEP would reduce average cost per available year, provided that the JSF procurement unit cost exceeded \$67.9 million.

Use of the average-cost-per-available-year metric consistently produces results that favor SLEP over a replacement aircraft, but to a lesser extent than using the cost-minimization approach. With our baseline parameters of a \$26 million E/F SLEP that adds ten years of service instead of immediately purchasing an \$80 million JSF, the model choice is moot. Both methodologies favor the SLEP. But the average-cost-per-available-year metric, because it considers aircraft availability, has a lower SLEP cost cutoff (\$30.9 million versus \$33.6 million), a greater SLEP-year cutoff (nine extra years versus eight), and a greater JSF procurement unit cost cutoff (\$67.9 million versus \$62.1 million).

Net Benefit–Maximization Methodology

Additional issues arise if the Navy wishes to maximize annualized net benefits (z_N or z_R) rather than minimize annualized costs (x_N or x_R) or average cost per available year. Most immediately, the Navy would need to know the values of the availability valuation parameters k_N and k_R.

We assume that the net benefit of buying JSFs is positive; the issue is whether to do so before or after E/F SLEPs. If

$$\sum_{t=0}^{Y_N} \frac{k_N \times a_N(t) - c_N(t)}{(1+i)^t} > 0,$$

it follows that

$$\sum_{t=0}^{30} \frac{k_{JSF} \times a_{JSF}(t)}{1.027^t} > \$148.38M,$$

where \$148.38 million is the discounted sum of estimated life-cycle procurement, maintenance, and modification costs in Table 4.3 and assuming that the Navy uses the OMB's prescribed 2.7-percent real interest rate to make investment decisions. Further, we can use the column of estimated composite availability levels as the $a_{JSF}(t)$ values, so we end up with the inequality

$$k_{JSF} > \frac{\$148.38M}{1{,}133.6} = \$130{,}891.$$

The value of k_{JSF} must be greater than \$130,891 in order for the JSF to have positive net benefit.

As noted in Chapter Three,

$$\sum_{t=1}^{Y_N} \frac{z_N}{(1+i)^t} = \sum_{t=0}^{Y_N} \frac{k_N \times a_N(t) - c_N(t)}{(1+i)^t}.$$

Hence,

$$\sum_{t=1}^{30} \frac{z_{JSF}}{1.027^t} = 1{,}133.6 \times k_{JSF} - \$148.38M,$$

so the annualized net benefit of a JSF would be $z_{JSF} = 55.6 \times k_{JSF} - \$7.28M$.

While \$130,891 forms a lower bound on k_{JSF}, there is additionally an upper bound on its value. In particular, referring to Table 4.3, we assume that it is appropriate to operate a JSF for 30 years rather than retiring it sooner (and, by assumption, replacing it with another JSF). If the Navy replaced the JSF sooner, it would get greater availability levels. If the Navy valued aircraft availability highly enough, it would not tolerate the late years of assumed 47.7-percent composite availability.

Hence, it must be that the net benefit from continuing JSF operation in year 30 is greater than buying a replacement aircraft that year. Therefore, it must be that $47.7 \times k_{JSF} - 5.23M > 55.6 \times k_{JSF} - 7.28M$, implying that we must have $\$257{,}721 > k_{JSF}$. We must have k_{JSF} greater than \$130,981 to give the JSF a positive present value, but less than \$257,721 to make operating the JSF out to age 30 preferable to retiring it sooner.

Just as we can express the annualized benefit of a JSF as $z_{JSF} = 55.6 \times k_{JSF} - 7.28M$, we can likewise express the annualized benefit of a prospective E/F SLEP as a function of $k_{E/F}$. For instance, using the parameters in Table 4.2, we find that the annualized benefit of a ten-year E/F SLEP would be $z_{E/F} = 53.2 \times k_{E/F} - 6.40M$.

There are three conditions that must hold for a ten-year E/F SLEP to be worth undertaking:

1. It must have a positive net present value, so $53.2 \times k_{E/F} - 6.40M > 0$.
2. It must have greater annualized net benefit than buying a JSF, so $53.2 \times k_{E/F} - 6.40M > 55.6 \times k_{JSF} - 7.28M$.
3. It is preferable to operate the SLEPed E/F in its tenth year post-SLEP rather than to switch to a JSF before that year. Hence, $47.7 \times k_{E/F} - 3.89M > 55.6 \times k_{JSF} - 7.28M$.

Given that we assume that $55.6 \times k_{JSF} - 7.28M \geq 0$, condition 2 subsumes condition 1. If k_{JSF} is at its minimum value of \$130,981, condition 2 holds if $k_{E/F} > \$120{,}246$. Condition 3 is nonbinding with $k_{E/F} > \$89{,}097$. If k_{JSF} is at its maximum value of \$257,721, condition 2 holds if $k_{E/F} > \$252{,}744$, while condition 3 again fails to bind at $k_{E/F} > \$237{,}036$.

Suppose the Navy valued SLEPed E/Fs and JSFs equivalently. Then $k_{JSF} = k_{E/F} = k$, so condition 2 would simplify to $\$369{,}066 > k$, while condition 3 would simplify to $\$382{,}003 > k$. Given that we know that $k = k_{JSF} \in [130981, 257721]$, each of these constraints must hold. If the two alternative aircraft are valued at parity, doing the \$26 million ten-year E/F SLEP would be optimal. This is the same finding we got through cost minimization and minimization of average cost per available year.

Continuing with $k = k_{JSF} = k_{E/F}$ parity, let the cost of the ten-year E/F SLEP be some value c_{SLEP} (that may be more or less than \$26 million). Then the life-cycle cost of the ten-year E/F SLEP case would be $c_{SLEP} + 29.45M$, and the annualized benefit of the ten-year SLEP would be

$$z_{E/F} = 53.2 \times k - 3.40M - \frac{c_{SLEP}}{8.66}.$$

The ten-year E/F SLEP will have a greater net benefit than buying a JSF if

$$53.2 \times k - 3.40M - \frac{c_{SLEP}}{8.66} > 55.6 \times k - 7.28M.$$

This inequality holds if

$$3.88M > 2.4 \times k + \frac{c_{SLEP}}{8.66}.$$

At the minimum possible k value of \$130,981, the c_{SLEP} cutoff would be \$30.9 million. At the maximum possible k value of \$257,721, the c_{SLEP} cutoff would be at \$28.2 million. By contrast, in Figure 5.1, we found that a SLEP that adds ten years of operation to an E/F can cost up to about \$33.6 million and still reduce Navy costs. The difference between Figure 5.1's result and our result here is that the net-benefit formulation puts additional positive weight on the high availability years associated with a new JSF. The higher the k value, the more positive weight put on those years, and the lower the SLEP cost cutoff for the ten-year SLEP to be worth doing. Note, however, that even our highest possible parity k case finds a \$26 million ten-year SLEP to be worthwhile.

We can run a similar calculation analyzing different JSF procurement unit costs. Let c_{JSF} denote a JSF's procurement unit cost. The annualized net benefit of a JSF would be

$$z_{JSF} = 55.6 \times k - 3.35M - \frac{c_{JSF}}{20.38}.$$

Then a \$26 million ten-year E/F SLEP will have a greater net benefit than buying a JSF if

$$53.2 \times k - 6.40M > 55.6 \times k - 3.35M - \frac{c_{JSF}}{20.38}.$$

This inequality holds if $c_{JSF} > 48.5 \times k + 62.1M$. At the minimum possible k value of \$130,981, the c_{JSF} cutoff would be at \$68.4 million. At the maximum possible k value of \$257,721, the c_{JSF} cutoff would be at \$74.6 million. By contrast, in Figure 5.3, we found a cutoff of about \$62.1 million (corresponding, not coincidentally, to the $k = 0$ value in the inequality just presented). When aircraft valuation is considered, we find that the JSF's procurement unit cost can be higher before an E/F SLEP is preferred. Again, however, our maximum c_{JSF} cutoff of \$74.6 million is still below our baseline \$80 million JSF procurement unit cost assumption.

Results are more ambiguous if there is uncertainty as to the number of extra years of operation emanating from a \$26-million SLEP. As noted, with a ten-year, \$26-million SLEP and k equality, the SLEP is worthwhile if $53.2 \times k - 6.40M > 55.6 \times k - 7.28M$, which simplifies to $\$369{,}066 > k$. Since we know that $k \in [130981, 257721]$, this inequality holds and a ten-year \$26 million SLEP is worthwhile.

The comparable inequality for a nine-year, $26-million SLEP is

$$47.9 \times k - 6.15M > 55.6 \times k - 7.28M,$$

which simplifies to $\$146{,}761 > k$. This value is within our *k* uncertainty range, so a nine-year, $26-million SLEP might or might not be worthwhile.

An eight-year, $26-million SLEP implies $42.6 \times k - 5.88M > 55.6 \times k - 7.28M$, which holds if $\$107{,}569 > k$. This inequality cannot hold given our constraints on *k*.

Hence, assuming *k* equality, a $26-million SLEP is worthwhile if it provides ten or more years of additional operation, is not worthwhile if it provides eight or fewer years, and might or might not be worthwhile if it provides nine extra years of operation.

Another scenario that can turn decisions against E/F SLEPs is when JSF availability is more highly valued than SLEPed E/F availability. Let

$$k_{JSF} = (1+d)k_{E/F}$$

or

$$k_{E/F} = \frac{k_{JSF}}{1+d}$$

with $d > 0$. A ten-year, $26-million E/F SLEP is preferred if

$$\frac{53.2}{1+d} \times k_{JSF} - 6.40M > 55.6 \times k_{JSF} - 7.28M$$

or

$$\frac{0.88M - 2.4 \times k_{JSF}}{55.6 \times k_{JSF} - 0.88M} > d.$$

At the minimum k_{JSF} value of $130,891, the *d* cutoff is 0.0885. At the maximum k_{JSF} value of $257,721, the *d* cutoff is 0.0197. Values of *d* larger than these cutoffs would have the ten-year, $26-million E/F SLEP no longer be chosen. In other words, if the Navy puts even a 2-percent greater value on JSF availability than on SLEPed E/F availability, it is possible that our baseline finding in favor of a ten-year, $26-million E/F SLEP would reverse. However, one view we heard is that SLEPed E/Fs and JSFs should be valued at parity because the Navy's most important objective is having enough fighter aircraft for its carriers.[1]

[1] There is a considerable literature on the "fighter gap," including Arthur and Eveker (2009), O'Rourke (2009), Congressional Budget Office (2010), and Tilghman (2010). If one is concerned about the "fighter gap," the central issue is having enough fighter aircraft to fill aircraft carriers with a lesser focus on what type of fighter aircraft. On the other hand, Shalal-Esa (2010) quotes Secretary of Defense Robert Gates on the possibility that F-35 capabilities might imply that legacy aircraft will not need to be replaced one-for-one.

Discussion

We have presented three different methodologies to evaluate the desirability of F/A-18E/F SLEPs. With our best estimates of relevant parameters (e.g., $26 million for a ten-year E/F SLEP versus buying an $80-million JSF), all three methodologies favor undertaking the E/F SLEP (see Table 5.1).

There are many uncertain parameters, however. For instance, given that C/D SLEPs have yet to be undertaken, we must attribute considerable uncertainty to the $26-million E/F SLEP, both in terms of its price tag and the extra years of operation it would provide. That said, the $80-million JSF procurement unit cost estimate could be an optimistic floor. If the JSF gets more expensive, there would be margin for an E/F SLEP to cost more than enumerated in Table 5.1 yet still be worthwhile.

The net benefit maximization's *d* parameter suggests where preference for a new JSF is most likely. If the Navy attributes even moderately greater military value to a JSF than to a SLEPed E/F, our pro-SLEP findings can reverse. With equal valuation of JSFs and SLEPed E/Fs, the available evidence favors SLEPs. The Navy must therefore examine the extent to which it might put greater military value on JSFs. Such an inquiry is beyond the scope of this analysis.

Table 5.1
Different Threshold Findings from Different Methodologies

Methodology	Maximum Cost of an E/F SLEP While Still Being Worth Doing ($ millions)	Maximum Value of JSF Procurement Unit Cost While JSF Is Still Preferred ($ millions)	Minimum Years for $26 Million SLEP to Be Worthwhile
Cost minimization	33.6	62.1	8
Average-cost-per-available-year minimization	30.9	67.9	9
Net benefit maximization	28.2–30.9	68.4–74.6	9 or 10

CHAPTER SIX

Conclusions

This report has applied three different methodologies (cost minimization, minimization of average cost per available year, and net-benefit maximization) to assessment of the desirability of an F/A-18E/F SLEP.

Minimization of average cost per available year and net-benefit maximization consider aircraft availability levels. Cost minimization does not. If we think new aircraft will generally have greater availability levels than repaired aircraft, the average-cost-per-available-year metric will tend to be lower for a new aircraft than for a SLEPed aircraft, thereby suggesting more value for new aircraft than that suggested by the constant availability cost-minimization methodology. The net-benefit maximization algorithm similarly tends to favor new aircraft more than pure cost minimization.

While the average-cost-per-available-year metric is tractable and intuitive in how it adjusts for changing availability, it is not a standard, agreed-upon metric. One must accept a priori that average cost per available year is the right metric to minimize. The net-benefit maximization approach assumes that Navy valuation of an aircraft increases linearly in $a(t)$. As noted at the end of Chapter Three, calculations would begin anew if a different net-benefit function, such as $p \times \sqrt{a(t)} - c(t)$, were asserted.

We conclude that an analysis of repair-replace decisionmaking should start with cost-minimization calculations. Most notably, if a new aircraft is found to be less costly, there is little need for additional calculations, assuming the typical result of newer aircraft having greater availability and capability levels. If, on the other hand, the repair approach is found to be less costly, we then recommend explorations of both net-benefit maximization and average-cost-per-available-year minimization to see how much additional availability or capability and the values placed on them it would take to offset the higher cost in decisionmaking.

Of course, if new aircraft are sizably preferred on military grounds, there is no need for the types of calculations presented in this report.

APPENDIX A

An Analysis of Carrier Onboard Delivery Options

Following up on the methodologies we developed to assess the desirability of F/A-18E/F SLEPs, the Navy also asked RAND to analyze carrier onboard delivery (COD) options. We discuss our findings in this appendix.

The Navy's primary COD aircraft is the C-2A. The C-2 design entered Navy service in the 1960s. In the 1980s, the original C-2 fleet was retired and replaced by the "re-procured C-2A" or C-2(R) (Younossi et al., 2004). Confusingly, these 1980s vintage aircraft are still referred to as C-2As, notwithstanding their being a second generation of this aircraft.

The C-2A fleet is currently wrapping up a series of maintenance actions, including a $1.6 million-per-aircraft SLEP,[1] a $2.5 million-per-aircraft rewiring, a $300,000-per-aircraft cockpit upgrade, and a $200,000-per-aircraft propeller modification.

There is considerable uncertainty as to how much longer the C-2A fleet will last after the completion of this suite of maintenance actions. Pessimistic projections suggest that the aircraft might need to be replaced or SLEPed again before 2020; more-optimistic projections suggest that the fleet will last until late in the 2020 decade.

Figure A.1 shows the C-2A fleet's quarterly EIS rates and MC rates dating back to the early 1990s. These data come from the Navy's DECKPLATE and AIRRS data systems.

Typically, about 70 percent of C-2As have been fleet possessed (the EIS rate) at a point in time, with those fleet-possessed aircraft averaging about a 60-percent MC rate. MC rates are tallied only from fleet-possessed aircraft.

In Figure A.2, we multiply the two data sets in Figure A.1 to derive the C-2A fleet's quarterly composite availability rates as a function of average fleet age.

The C-2A has had low and moderately downward-trending composite availability rates. The typical C-2A composite availability rate has been in the low 40-percent range. Regression analysis suggested a statistically significant downward drift of about 1 percent per year of age in Figure A.2.

As in the body of this report, we need to project the C-2A's future annual O&S costs as well as life-cycle O&S costs of any replacement aircraft.

In accord with AIR 4.2.2 (2009), we assume that a 21-year-old C-2A will have annual O&S costs per aircraft around $8.4 million. We further assume that the maintenance components of O&S costs (unit-level consumption, intermediate maintenance, depot maintenance,

[1] There is a semantic problem in comparing F/A-18E/F issues and C-2A issues. In the F/A-18 context, the term *SLEP* is used to describe an entire set of maintenance actions. By contrast, the C-2A program office uses the term to refer to only a portion of the work that was recently performed on the fleet. We use the term *SLEP* in the more expansive way in which the F/A-18E/F program office uses the term. So, in our nomenclature, all the recent C-2A maintenance activities would be referred to as a SLEP of roughly $4.6 million per aircraft.

Figure A.1
C-2A Equipment in Service and Mission-Capable Rates over Time

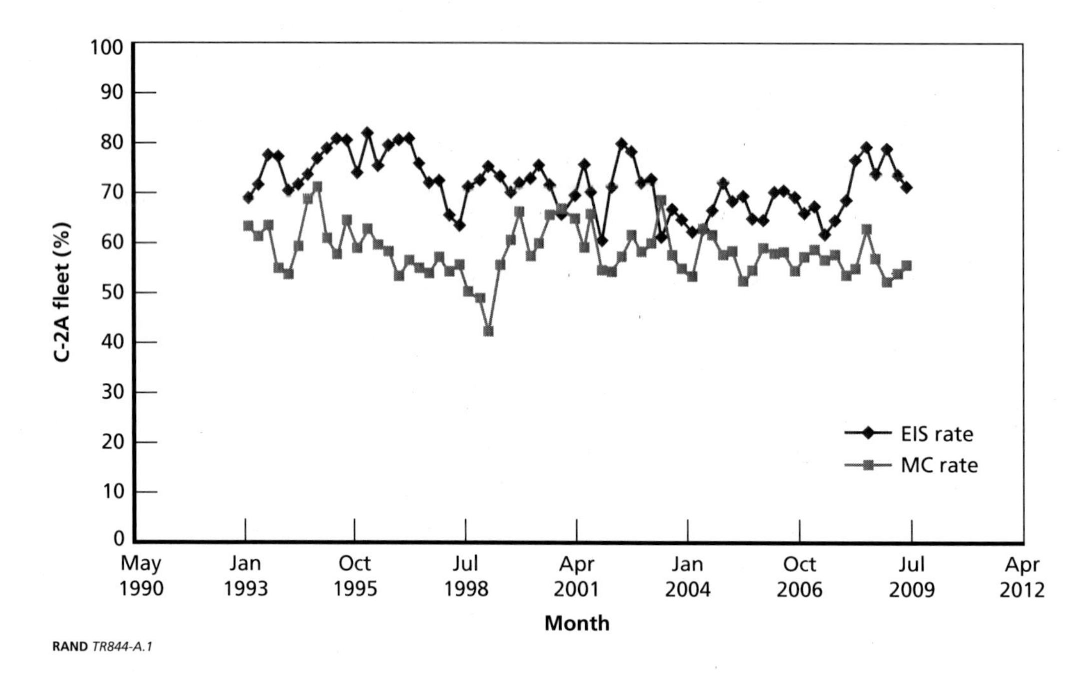

RAND *TR844-A.1*

Figure A.2
C-2A Composite Availability Rates, as a Function of Average Fleet Age

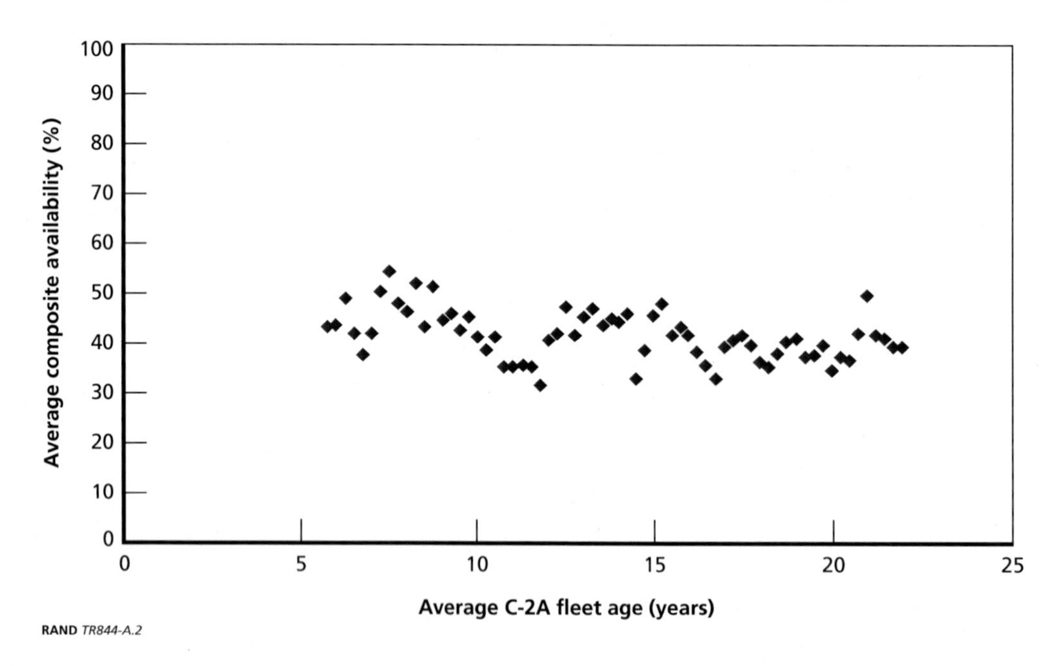

RAND *TR844-A.2*

and sustaining support) will increase at 3 percent per year in real terms but that mission personnel and indirect support expenditures will be constant in real terms.

If the C-2A receives another SLEP, we assume that its annual O&S costs will roll back accordingly, e.g., a ten-year SLEP on a C-2A will result in an aircraft with the O&S cost and availability level of an aircraft 11 years younger (with the extra year accounting for the assumed duration of the SLEP). All of these assumptions are parallel to those we made in our F/A-18E/F analyses.

A prospective C-2A replacement aircraft is estimated to have considerable costs $\left(c_N(0)\right)$. Data we received from the program office indicated that a replacement COD aircraft would cost $3.16 billion in FY 2009 dollars for development (including five aircraft) and another $5.37 billion for production of 40 additional aircraft. In total, 45 aircraft would be produced, with an average cost of about $190 million per aircraft. We assume that such a replacement aircraft would be operated for 30 years $\left(Y_N = 30\right)$.

Analysis of C-2A Service Life Extension Program Desirability

The $4.6 million worth of maintenance actions that are currently ongoing in the C-2A fleet were a terrific value measured against a $190 million replacement, we find. In Figure A.3, we compare the average cost per available year associated with a $4.6-million SLEP against a new $190-million aircraft. The average-cost-per-available-year metric is better than a new aircraft irrespective of how many years of extra operation the SLEP provides.

One attains the same result using Chapter Two's cost-per-year metric that does not give the new aircraft credit for greater availability rates. One Navy expert with whom we spoke opined that this SLEP has been a "raging success." Figure A.3 is consistent with that assertion.

In Figure A.4, we estimate how much the Navy might be willing to pay for a ten-year C-2A SLEP. Using the average-cost-per-available-year metric, we find the breakeven to be around $62 million. The cost-per-year metric has an even greater breakeven, around $70 million. As discussed in the body of this report, we expect the cost-per-year metric to be more favorable to aircraft repair than the average-cost-per-available-year metric because the former does not account for increased availability associated with new aircraft.

Figure A.3
Assessment of the Desirability of Recent C-2A Maintenance Actions

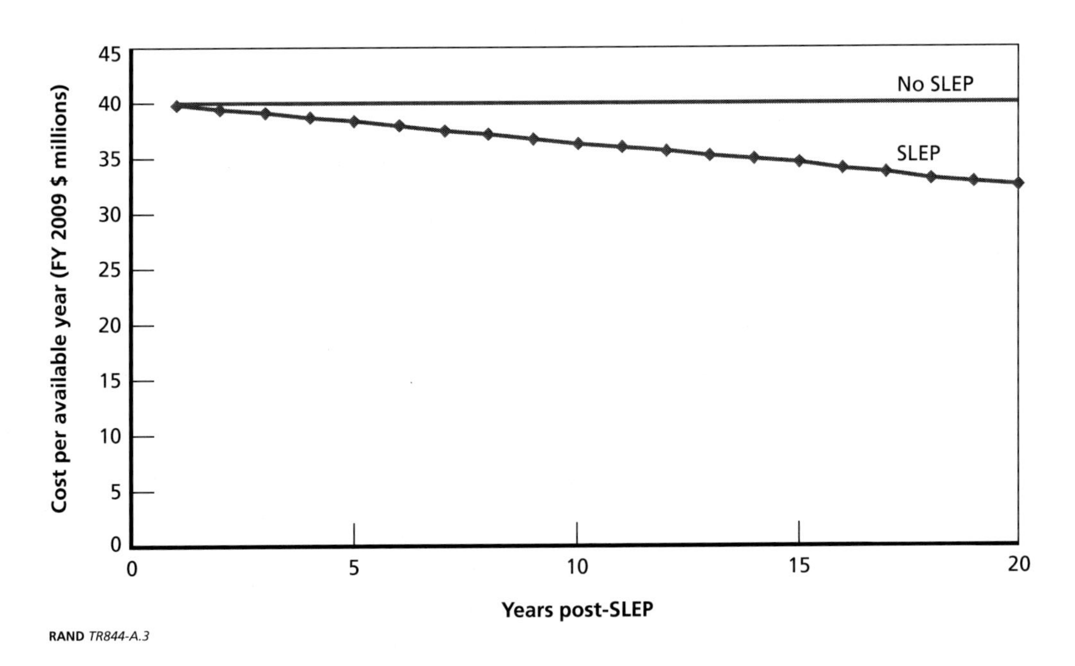

RAND *TR844-A.3*

Figure A.4
Maximum Willingness to Pay for a Ten-Year C-2A Service Life Extension Program

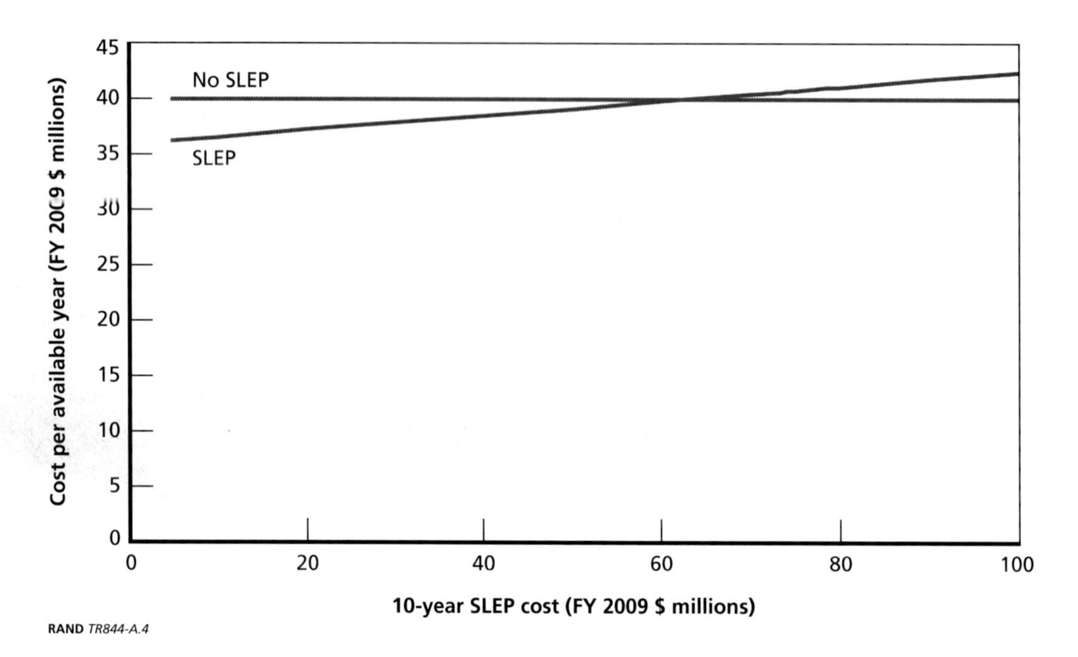

RAND *TR844-A.4*

Compared to the $4.6 million per aircraft the Navy is spending on the current set of C-2A maintenance actions, there is considerable room for more-expensive repairs while having them still be preferred to a new COD aircraft.

Figure A.5 compares SLEPs, adding between one and 20 years of operation to the C-2A fleet. Not surprisingly, the Navy's maximum willingness to pay for a SLEP increases in the number of additional years of service provided.

The cost-per-year metric that does not give credit for new aircraft having greater availability levels results in greater willingness to pay for a SLEP than the average-cost-per-available-year metric that does give new aircraft credit for additional availability.

It is also no surprise that, when the prospective replacement aircraft is more expensive, the Navy's maximum willingness to pay for a ten-year SLEP increases, as presented in Figure A.6.

Using the average-cost-per-available-year metric (the lower line), the Navy would be willing to pay up to about $50 million for a ten-year SLEP if the replacement aircraft had a unit cost of $160 million. But if the replacement aircraft had a unit cost of $220 million, the Navy would be willing to pay up to about $75 million for the same ten-year SLEP, we find.

We have superimposed the vertical axis in Figure A.6 at our baseline assumption of a replacement aircraft having a unit cost of about $190 million.

Ultimately, the Navy has four options. It can continue to maintain the C-2A fleet, including requisite SLEPs. It can replace the C-2As with new aircraft. It can find other aircraft or ships that can fulfill C-2A missions. Or it can live without fulfillment of current C-2A missions.

We think that it would be valuable to identify prospective COD substitute aircraft. Of course, it would be attractive if there were a prospective replacement that cost less than $190 million per aircraft. Also, assuming that another SLEP will be undertaken on the C-2A fleet, the number of C-2As not in depot-level maintenance may fall below requisite levels. In this case, there would be a need to supplement the COD fleet, if only temporarily.

Figure A.5
Maximum Willingness to Pay for a Service Life Extension Program, as a Function of Additional Years of Service Provided

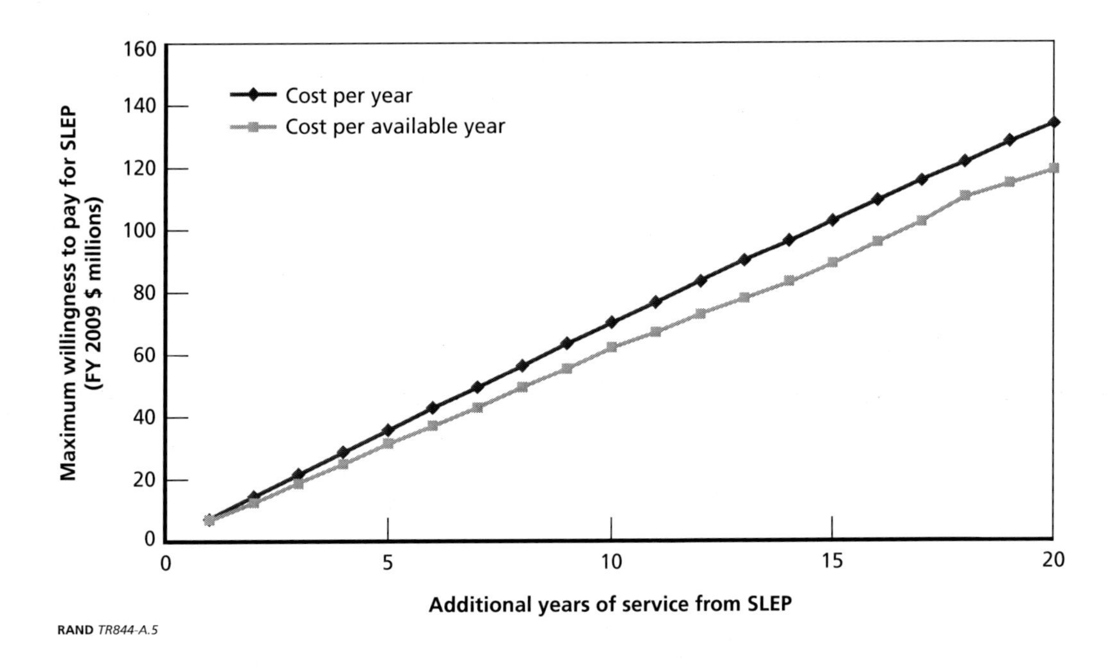

RAND *TR844-A.5*

Figure A.6
Maximum Willingness to Pay for a Ten-Year Service Life Extension Program, as a Function of Replacement-Aircraft Cost

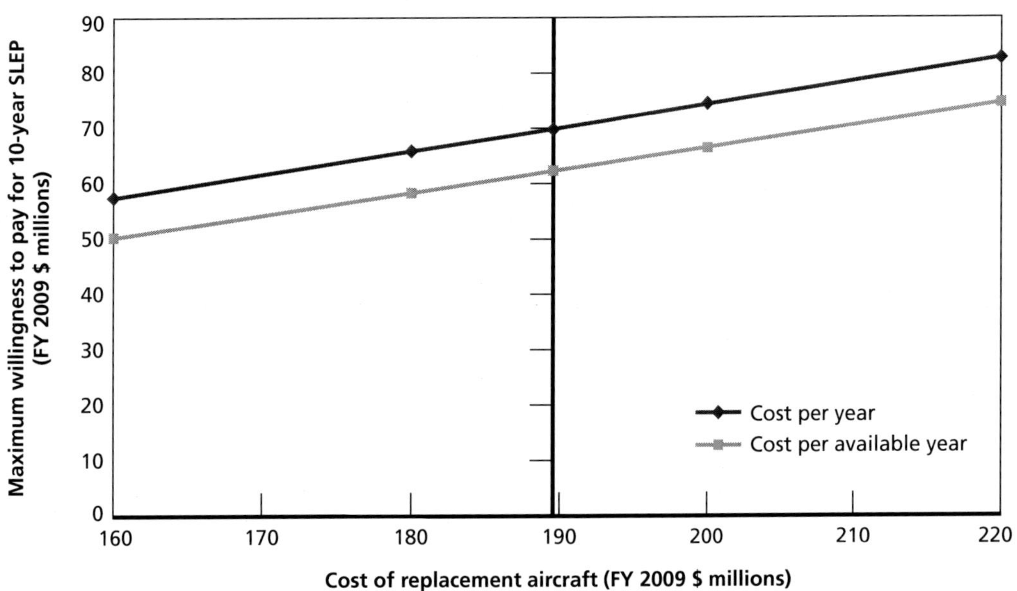

RAND *TR844-A.6*

APPENDIX B

Net Present Value Versus Annuitized Value in SLEP Analysis

Standard project evaluation and capital budgeting analysis use net present value as the criterion for comparing alternative investment options. OMB Circular A-94 prescribes its use in government cost-benefit analysis. This appendix demonstrates the equivalence between a standard net present value approach and the approach used in this report.

In Chapter Two, we defined annuitized values x_N and x_R, where

$$x_N = \frac{\sum_{t=0}^{Y_N} \frac{c_N(t)}{(1+i)^t}}{\sum_{t=1}^{Y_N} \frac{1}{(1+i)^t}}$$

and

$$x_R = \frac{\sum_{t=0}^{Y_R} \frac{c_R(t)}{(1+i)^t}}{\sum_{t=1}^{Y_R} \frac{1}{(1+i)^t}}.$$

We asserted that the Navy wants to undertake a Y_R-year SLEP and postpone buying the new aircraft if and only if $x_R < x_N$.

To calculate net present value, V_N, we assume that the new, replacement aircraft would be replaced by its clone indefinitely into the future, so that

$$V_N = \sum_{t=1}^{\infty} \frac{x_N}{(1+i)^t}.$$

By contrast, we assume that the SLEPed aircraft would be operated for Y_R years then be replaced by the succession of new replacement clones so the net present value of the repaired aircraft, V_R, would be

$$V_R = \sum_{t=0}^{Y_R} \frac{c_R(t)}{(1+i)^t} + \frac{V_N}{(1+i)^{Y_R}}.$$

The Navy wants to undertake the SLEP if and only if $V_R < V_N$ or

$$\sum_{t=0}^{Y_R} \frac{c_R(t)}{(1+i)^t} + \frac{V_N}{(1+i)^{Y_R}} < V_N.$$

As noted,

$$V_N = \sum_{t=1}^{\infty} \frac{x_N}{(1+i)^t},$$

so

$$V_N - \frac{V_N}{(1+i)^{Y_R}} = \frac{x_N}{(1+i)} + \frac{x_N}{(1+i)^2} + \ldots + \frac{x_N}{(1+i)^{Y_R}} + \frac{x_N}{(1+i)^{Y_R+1}} + \ldots - \frac{x_N}{(1+i)^{Y_R+1}} - \frac{x_N}{(1+i)^{Y_R+2}} - \ldots,$$

which simplifies to

$$\sum_{t=1}^{Y_R} \frac{x_N}{(1+i)^t}.$$

Thus, our net present value SLEP inequality is

$$\sum_{t=0}^{Y_R} \frac{c_R(t)}{(1+i)^t} < \sum_{t=1}^{Y_R} \frac{x_N}{(1+i)^t}.$$

Dividing both sides of the inequality by

$$\sum_{t=1}^{Y_R} \frac{1}{(1+i)^t},$$

we get $x_R < x_N$. For this type of problem, a cost analysis framed in terms of annuitized values yields the same policy implications as a cost analysis framed in terms of net present value. Greenfield and Persselin (2002) applied a net present value approach to a military aircraft repair-replace problem.

References

AIR 4.2.2, *C-2A Program Operating and Support Cost Analysis FY2009*, September 2009.

Arthur, David, and Kevin Eveker, *Alternatives for Modernizing U.S. Fighter Forces*, Washington, D.C.: Congressional Budget Office, May 2009. As of September 28, 2010:
http://www.cbo.gov/ftpdocs/101xx/doc10113/05-13-FighterForces.pdf

Bureau of Economic Analysis, "National Income and Product Accounts Table: Table 1.1.9 Implicit Price Deflators for Gross Domestic Product," last revised October 29, 2009. As of October 31, 2009:
http://www.bea.gov/national/nipaweb/SelectTable.asp?Selected=N

Congressional Budget Office, *Strategies for Maintaining the Navy's and Marine Corps' Inventories of Fighter Aircraft*, Washington, D.C., May 2010. As of September 28, 2010:
http://purl.access.gpo.gov/GPO/LPS122569

Defense Acquisition University, "Cost Terms," *ACQuipedia*, undated web page. As of March 3, 2010:
https://acc.dau.mil/CommunityBrowser.aspx?id=243016

Dixon, Matthew C., *The Maintenance Costs of Aging Aircraft: Insights from Commercial Aviation*, Santa Monica, Calif.: RAND Corporation, MG-486-AF, 2006. As of September 28, 2010:
http://www.rand.org/pubs/monographs/MG486/

Drew, Christopher, "U.S. May Add Money to Program for F-35 Jet," *New York Times*, November 21, 2009, p. B4. As of September 28, 2010:
http://www.nytimes.com/2009/11/21/business/21plane.html

Greenfield, Victoria A., and David Persselin, *An Economic Framework for Evaluating Military Aircraft Replacement*, Santa Monica, Calif.: RAND Corporation, MR-1489-AF, 2002. As of September 28, 2010:
http://www.rand.org/pubs/monograph_reports/MR1489/

Keating, Edward G., and Matthew C. Dixon, *Investigating Optimal Replacement of Aging Air Force Systems*, Santa Monica, Calif.: RAND Corporation, MR-1763-AF, 2003. As of September 28, 2010:
http://www.rand.org/pubs/monograph_reports/MR1763/

Keating, Edward G., Don Snyder, Matthew C. Dixon, and Elvira N. Loredo, *Aging Aircraft Repair-Replacement Decisions with Depot-Level Capacity as a Policy Choice Variable*, Santa Monica, Calif.: RAND Corporation, MG-241-AF, 2005. As of September 28, 2010:
http://www.rand.org/pubs/monographs/MG241/

Kiley, Gregory T., and John Skeen, *The Effects of Aging on the Costs of Operating and Maintaining Military Equipment*, Washington, D.C.: Congressional Budget Office, August 2001. As of September 28, 2010:
http://www.cbo.gov/doc.cfm?index=2982

Office of Management and Budget, *Guidelines and Discount Rates for Benefit-Cost Analysis of Federal Programs*, Washington, D.C., Circular A-94, revised January 22, 2002. As of September 28, 2010:
http://purl.access.gpo.gov/GPO/LPS46031

———, *Discount Rates for Cost-Effectiveness, Lease Purchase, and Related Analyses*, Washington, D.C., Circular A-94, Appendix C, December 2009. As of December 19, 2009:
http://www.whitehouse.gov/omb/circulars_a094_a94_appx-c/

OMB—*See* Office of Management and Budget.

O'Rourke, Ronald, *Navy F/A-18E/F and EA-18G Aircraft Procurement and Strike Fighter Shortfall: Background and Issues for Congress*, Washington, D.C.: Congressional Research Service, October 21, 2009. As of September 28, 2010:
http://handle.dtic.mil/100.2/ADA509770

Pyles, Raymond A., *Aging Aircraft: USAF Workload and Material Consumption Life Cycle Patterns*, Santa Monica, Calif.: RAND Corporation, MR-1641-AF, 2003. As of September 28, 2010:
http://www.rand.org/pubs/monograph_reports/MR1641/

Scully, Megan, "Republican Renews Bid for Multiyear F-18 Deal," *Government Executive.com*, February 22, 2010. As of March 1, 2010:
http://www.govexec.com/dailyfed/0210/022210cdpm1.htm

Shalal-Esa, Andrea, "Gates Defends F-35, Rejects Increase in F/A-18s," *Reuters.com*, February 3, 2010. As of February 4, 2010:
http://www.reuters.com/article/idUSTRE61304H20100204

Sherman, Jason, "Navy Concerned JSF Price Tag Could Squeeze Out Other Aviation Needs," *Inside the Pentagon*, January 14, 2010.

Tilghman, Andrew, "Gates: Fighter Gap Ignores Real-World Demand," *Navy Times*, June 8, 2010, p. 19. As of September 28, 2010:
http://www.navytimes.com/news/2010/06/navy_fighter_gap_060710w/

U.S. Navy, "Fact File: F/A-18 Hornet Strike Fighter," last updated May 26, 2009. As of October 31, 2009:
http://www.navy.mil/navydata/fact_display.asp?cid=1100&tid=1200&ct=1

Younossi, Obaid, Kevin Brancato, Fred Timson, and Jerry Sollinger, *Starting Over: Technical, Schedule, and Cost Issues Involved with Restarting C-2 Production*, Santa Monica, Calif.: RAND Corporation, 2004. Not available to the general public.